# THE HARRIS ORTHOPAEDIC LABORATORY

## @

## *The Mass General*

William H. Harris, MD, DSc

# CONTENTS

# LIST OF FIGURES
# AND TABLES

## FIGURES

## TABLES

# PREAMBLE

## HIGHLIGHTS OF THE FIRST FIVE DECADES OF THE HARRIS ORTHOPAEDIC LABORATORY

- Initiation of the first attack on, followed by multiple contributions to, the reduction of fatal pulmonary embolism after hip surgery.
- Important contributions to improved understanding of the inherent linkage between bone resorption and formation.
- Major contributions to the success of human limb reimplantation.
- Key contributions to the new understanding of the etiology of osteoarthritis of the hip.
- Unique direct measurement of the pressure in human articular cartilage in vivo during activities of daily living.
- Major contributions to advances in total hip implant design and techniques for both cemented and cementless total hip replacements.
- Significant improvements for fixation of and the functioning of both cemented and cementless THR.
- A major role in unraveling the mysteries of the unique stealth disease, periprosthetic osteolysis.
- Creation of the new world's standard for hip simulators, crucial to advancing innovation in hip surgery design and materials.

- Initiation of and direction of the development of highly crosslinked polyethylene, an outstanding contribution to total hip and total knee surgery and a major advance in the elimination of periprosthetic osteolysis worldwide.
- Major innovations in creating reconstructive hip surgery for arthritis caused by developmental dysplasia and total developmental dislocation.
- Important contributions to the rigorous quantification of both the hip-centered outcomes of surgery of the adult hip and of the individual techniques of hip surgery.
- Creating the world's first hip registry which included full integration of digital radiographs.
- Established the capacity of highly crosslinked polyethylene to resist wear with sufficient power to safely permit the use of femoral heads greater than 32 mm in diameter, and thus effectively reducing the dislocation rate of metal-on-polyethylene THR.
- The combined publications from the Harris Orhtopaedic Laboratory from 1969-2017, exceeding 600 papers, were granted three Kappa Delta National Orthopaedic Research Awards, 10 Hip Society Awards, the Hap Paul Award three times, and had the highest number of citations in four different categories of orthopaedic surgery, the 100 classic papers in orthopaedic surgery, the 100 most cited papers in orthopaedic surgery, the 100 most cited papers in orthopaedic hip research and the 50 highest cited papers in hip and knee arthroplasty.

# DEDICATION

Dedication: Part I:
My Mentors and My Fellows

THIS BOOK IS DEDICATED TO two most valuable groups of people – my mentors and my Fellows. Although the title is short, the list is long. The list of mentors begins with my family, Mother and Dad, and brother Jack. Mother and Dad were two remarkable human beings whose life progression, beginning in modest circumstances in tiny towns in rural Kansas, exemplified compassion, quality and contribution. Mother instilled love and exhibited dedication to others, while Dad, in addition to those attributes, demanded rigor and quantification. Brother Jack, my guide and protector throughout the early formative years, taught me many of the fundamentals of living.

I would honor Evelyn Rainsford, my third grade teacher, Viola Wilt, a rigorous Latin teacher, and Miss Gingrich, who extolled sentence structure while making learning exciting.

Major players in my becoming further educated were "Unc Toby", an English teacher, Dave Chapman whose title claimed that he taught history but he really taught life, and George Hammer, the precise, demanding and exquisite physics teacher – all at the Mercersburg Academy. It was there that the sharp transition between "going to school" versus "becoming educated" was swiftly and forcefully accomplished.

My specific key mentors at Haverford College were chemistry professors, William Meldrum and Bill Cadbury but the non-specific key mentors were the extraordinary talented fellow students who comprised the student body.

In medicine and then in surgery at Penn strong leaders abounded. I. S. Ravdin, the Chief of Surgery and Julien Johnson, Chief of Thoracic Surgery, established intensive, demanding standards of care for the surgically sick, and were mentors of lasting influence.

In orthopaedics J.S. Barr, W.T. Green and Otto Aufranc – all for quite different reasons – proved to generate salient leadership, demonstrating remarkable skills and setting high standards, not only in the surgical aspects of musculoskeletal care but for the crucial human interactions in medicine.

And in musculoskeletal research, the names of Jennifer Jowsey in London and Melvin J. Glimcher at the MGH illustrated potent, but quite diverse leadership and mentorship. So too do the names of Edwin Salzman, Roman DeSanctis and Edward Merrill rise to the top as remarkable colleagues during these travels. And I was unusually blessed when it became time to relinquish the reins to be able to turn over everything I had built to two exquisite coinvestigators of unique contributions and distinction, fully capable of thrusting our combined quests further, Orhun Muratoglu and Henrik Malchau. And while more as a companion-mentor than a professor-mentor, Edith Weinberg's contributions to all of my work and to the formation of the Harris Orthopaedic Laboratory were both immense and essential at multiple levels, organizationally, scientifically and supportively.

Of all the multifaceted interactions along this 67-year adventure since obtaining the MD degree, none exceed the richness, depth, rewards and future impact of the interplay more than my Fellows, 100 strong. That they would be remarkably successful was fully assured before they came to the fellowship, simply because they comprised a fully proven, remarkably gifted, highly selected group. The raw and not-raw talent of the group was fully manifest from the start.

My joy was to work with these surgeons already possessing that talent. The fellowship created a huge opportunity for me. It allowed me to relish in one of my most favorite activities, molding very smart minds. Without exception, a favorite exercise of mine was to challenge accomplished students, to move from simply data transfer or the description of concepts to the creation of ideas and to the challenge of addressing the unknown. My most rewarding questioning within my Socratic approach to teaching was, basically, regardless of the topic, "what are you going to do when you don't know what to do?"

The fellowship held two other massive rewards for me, one "present" and one "future." The present reward was the capacity of the Fellows to multiply the depth and efficacy of our research programs. These activities were greatly augmented and expanded by the application of their skills, interests, commitments and contributions. Their role was pivotal.

And simultaneously, the educational input to them was often determinative subsequently, with many "offspring" laboratories and practices bearing the fingerprints of the Harris Orthopaedic Lab. The "future" reward lay in the extent, viability and productivity of these very capable individuals in their own right, once activated from the launchpad of the fellowship. This group of joint surgeons, educators, innovators and leaders has proven themselves many times over. And, even more so, they constitute a wealth of warm personal human interactions which augment my extended family.

Simple enumeration belies the depth, inspiration, challenges and zest that interactions with this remarkable group of unusual human beings has meant to me and to the progress of my lab over these five-and-a-half decades. To them all, a very meaningful and rich "Thanks".

Dedication: Part II:
One More Essential Ingredient:
Disciplined Dissatisfaction

And further, in a quite different setting, a portion of this dedication must focus on "disciplined dissatisfaction". Dissatisfaction cries out, "Why can't we do better than this! Is this the best there is? Why are we limited to this?"

Without that stimulus, little will change.

But the truly effective additive is the word "disciplined" in the phrase "disciplined dissatisfaction". Without the discipline to coerce the dissatisfaction into progress, the dissatisfaction is "but a hollow sounding bell." Real change requires both.

# GUIDING PHILOSOPHY

COMMON TO THE FORMATIVE PERIOD of many of the youth of my era, I was imbued with the pervasive commitment to project-oriented problem-solving. This orientation was certainly developmentally instilled if not genetically mandated. But ultimately its manifestation depended heavily on serial reinforcement, at home, at the Mercersburg Academy, at Haverford College, at the Perelman School of Medicine at the University of Pennsylvania and during orthopaedic training at Harvard.

The serial steps in generating my guiding philosophy were the selection of medicine as a career goal at age eight, an evolving decision toward orthopaedics over the years between ages 13 and 27, with the final commitment to a clinician-scientist career at the MGH with an essential focus on issues concerning the hip. As is always true, this outcome resulted from the interplay of motivation, change, chance and opportunity. Polio played a dual role, in both change and chance. The "change" was the elimination of that disease by vaccine when I had first entered orthopaedics and the "chance" was from obtaining a Polio Fellowship (National Infantile Paralysis Scholarship) to fund a year of post-residency study in London. The attraction to return to the MGH grew from knowing first-hand its unyielding commitment to excellence in both clinical care and to innovation, exemplified at that time by the recent opening of the unique Orthopaedic Research Laboratory.

Conceptually the commitment to project-driven problem-solving was further defined by two concepts which I followed and have given steadfastly as advice to the young ever since. The first is "Go where no one else wants to go". This admonition takes you off the beaten track where it is far less crowded, provides ample room to both expand and achieve, and provides vistas replete with great problems to solve.

The second is "Follow the problem wherever it leads." This concept mandates flexibility in the type of skills or attributes that the research embraces and prevents the effort from being locked into becoming an effort with just "one trick" available. This approach does contain more risks, for the hazards of assuming one can develop or find the diverse skills required are real. Nevertheless the success of this approach is clearly shown by the improbable success of an orthopaedic surgeon being able to reduce the vastly different, severe medical problems of fatal pulmonary emboli and periprosthetic osteolysis.

# THE CLINICIAN-SCIENTIST

PROGRESS IN THE MANAGEMENT OF human disease is a hallmark of the human adventure. Mainly a product of Western Civilization and largely observational and descriptive during the dawn of Western Civilization, the extraordinary accelerations of this key process and the resulting reduction of human suffering benefitted hugely from the Enlightenment and subsequent scientific revolutions.

Throughout the long history of medicine, the overwhelming majority of medical practitioners have dispensed their art, knowledge and skills by focusing what they considered to be valid understandings of disease and therapeutics directly on their patients in the hope that this would improve the outcome. Even if much of the perceived "valid" information was erroneous, the basic principle then, as today, was to apply their best perceived recommendation to each patient.

However admirable, by itself, that concept alone has the severe limitation that, per se, it does not <u>advance</u> medical care. To meet <u>that</u> specific and essential need, there must be a subset of practitioners who sense a separate responsibility, that of <u>improving</u> the understanding and/or the capabilities of medical care. They are represented, for example, by those early innovators who developed the techniques such as trephining the skull, as practiced in ancient civilizations. During the early millennia of recorded history, most of these changes which represented progress in medical care were observational, made by those who generated novel insights of the disease processes. Outstanding among them were the contributions of Hippocrates and his followers, whose work was both identified by and characterized by <u>observation and documentation</u>. From those two elements, conclusions were drawn which created the first steps toward modern medicine. Similarly Galen, nearly six hundred years later, followed parallel practices and significantly improved medicine predominantly on the basis of observations of disease itself.

Augmenting many of these advances were simply issues of common sense. Vesalius, in order to stem the flow of blood from the arteries during amputations, replaced the barbaric practice of cauterization of the wound by uniquely applying ligatures to the arteries, a major advance. Other changes, of course, did not represent progress but grew out of erroneous beliefs such as the concept that "bleeding" would represent treatment for many diseases.

Importantly, the understanding of disease was remarkably advanced by reversing the taboos against studying the human body directly, a change which was a by-product of the Renaissance. This placed the observational investigation of disease on a vastly enhanced basis not previously possible. Similarly, direct observations led to remarkable increases in the understanding of physiology, such as the demonstration by Harvey of the true nature of blood circulation.

But by far the most important single advance was the expansion of science, science of all types, beginning with the Renaissance, expanding first during the Enlightenment, and then progressively exploding further throughout the 20th and 21st centuries.

The application of scientific thought accelerated progress in medicine by huge increments, well beyond that made possible simply by observations. For example, while Jenner observed two different but related diseases, the smallpox and the cowpox, his concept went far beyond the simple observation of each disease. From his observations he originated a unique postulate, a relationship between the two and thus generated the idea of a vaccine. Such changes were not simply observations of disease itself but rather represented conceptual advances in the creation of entirely new forms of treatment and/or prophylaxis.

While Semmelweis also generated a huge step forward simply by insightfully observing disease, Koch innovated beyond just observations with his advanced concepts in terms of his postulates on bacterial infections. Likewise, Virchow's triad was not simply a matter of observing and documenting but rather applied new science to understand the observed process of the clotting of blood.

Increasingly as science independent of medicine made remarkable advances in so many other fields, it was possible to draw meaningful insights and improvement from these many diverse branches of science. Mendel's observations in botany on the sweetpea in terms of inheritance opened the entire realm of genetic influences, leading ultimately to the characterization of the genome.

Becquerel's unique basic science observations which revealed radioactivity ultimately led to the work of the Curies and subsequently to the extraordinary expansion of understanding of both physiology and disease through the applications of radioactive isotopes. Roentgen's observation based on the science of physics created not only the entirely new field of radiographic diagnosis, but also the radiotherapy of tumors.

Medicine borrowed the curious property of nitrous oxide to render individuals transiently insensible, leading to the recognition that ether could provide reversible anesthesia.

Extraordinary advances in optics also led to such innovation as the development of the electron microscope and beyond, a pivotal step in understanding the intricacies of cellular structure, contents and behavior.

Engineering contributed enormously not only through its role in the creation of the heart-lung pump, but across such advances as unique fracture management, and the development of total joint arthroplasty.

Pasteur's motto that "chance favors only the prepared mind" was marvelously demonstrated in that particular chance observation which led Fleming to pursue the curious behavior on bacterial culture plates which were contaminated by a certain penicillin mold. That example, along with the sulfanilamide drugs, served as the stimuli for the extraordinary explosion of the development of antibiotics.

The cytology of cellular toxicity spawned the entire field of chemotherapy.

And so, from earliest times through to today, medicine has always had as its prime responsibility the <u>delivery</u> to patients the optimum therapeutics, but, in addition, central to the ability to <u>advance</u> medicine have been those small percentage of practitioners and scientists who were driven by a commitment to <u>improve</u> care. Those who both practice patient care and advance the field by introducing science to that process have been designated clinician-scientists.

Thus, quite in addition to the pure clinician, the advances in the prevention, elimination, management and control of disease lie in the hands of these three extra groups, clinician-scientists who may be either primarily clinician-observers or clinicians who add science to the mix and the pure scientists. Some of the pure scientists make their full commitment to the medical area, and others, in pursuit of their own science, simultaneously develop remarkable advances which either extend into the management of medical disease or serendipitously advance medical causes.

In fact the lineage of faculty at the MGH advancing the understanding and solution to diseases of the hip follows this specific evolution, beginning with Henry Jacob Bigelow, a pure but very imaginative clinician, through Smith-Petersen, a most inventive clinician but one who was not designated as a clinician-scientist while functioning as one, to O.E. Aufranc, a skilled pure clinician, to my efforts which clearly represented those of a clinician-scientist. This sequence led to the leadership of the Harris Orthopaedic Laboratory today, which is now directed by a strong clinically-oriented pure scientist, a materials engineer, Orhun Muratoglu, Ph.D.

After finishing my orthopaedic residency training at the Children's Hospital of Boston and the MGH, I cast my lot to find a career that encompassed both science and medicine, that of the clinician-scientist. Behind that decision lie two questions. Why become a clinician-scientist? And what factors contributed to my arriving at the selection?

Driven by an eclectic array of multiple diverse features and influences as widespread as my father's rigor and belief in science, a swift and fruitful educational challenge at Mercersburg Academy, both the love of science and the moral fiber of the liberal education at Haverford College, the stimuli of demanding role models of the faculty at Penn Medical School and a deeply-driven approach that constantly asked "why can't we do better than this?", the choice for me was easy.

The clinician-scientist role combined the richness and challenges of patient care, surgery, and the hands-on management of human disease with the insistent demands, complexities and rewards of advances in medically-oriented biology and engineering.

As the disruption of World War II subsided, orthopaedic departments throughout the United States and more slowly throughout the rest of the world, gradually reconstituted their staff and reoriented their efforts. As an unintended consequence,the wartime experience had led to major advances in many vital aspects of musculoskeletal care. Certainly the management of trauma had been vastly improved, enforced upon orthopaedics by the necessities of war. In addition, multiple other advances had occurred, not the least of which included the development of penicillin, improvements in fluid balance, advances in blood transfusions, and aggressive management of specific fractures, exemplified by the femoral intramedullary rod.

Still, as the academic units were re-formed, the principal model was the hoary form dating back to antiquity, with two general categories of faculty members, pure clinicians, and a small number of clinician-scientists. All these caregivers were involved in patient care but the small subset of clinician-scientists, in addition to the patient care, advanced orthopaedics using the traditional form of generating new insights primarily by techniques of observation and induction.

But, specifically, where to cast my lot?

The decision about aiming for medicine had been easy and early. My admiration for my father, a general practitioner initially who became a radiologist – and for his love of being a physician – made that decision quasi-congenital. And then as I worked all manner of odd jobs at the hospital, the idea of becoming an orthopaedic surgeon grew progressively, despite the common fascination that I shared with many medical students of being infatuated with each specialty among the clinical entities during every succeeding rotation. Each facet of patient care has compellingly attractive allure. Through it all, orthopaedic surgery had won out.

However, the massive explosion of basic science during and following World War II influenced much of the research on musculoskeletal disease, particularly in selected academic orthopaedic and medical departments. While this explosion commonly took the form of a unique change in the clinician-scientist profile, namely faculty members who simultaneously devoted some of their efforts to basic research and also played a role in clinical practice, beyond that were the full-time or nearly full-time basic scientists, attacking the problems of fundamental issues in musculoskeletal physiology and disease. Now research activities sprung up in such basic fields such as adverse reactions to metals, the understanding and prevention of osteoporosis, molecular biology, genetics, etc. These aspects represented a major departure from the past.

The specific steps following that decision were both easy and hard. The decision about the clinical area to focus upon was as easy as that for the the research area was hard.

Within the overall umbrella of all medical progress, one subset – surgical progress – has its own specific enabling and also limiting characteristics. As a limitation, for example, it is likely to remain the last bastion of medical therapy that exists as a cottage industry – remaining a "one on one" or person to person process well after "Watson" and other big data processes and processors have strongly modified much of the rest of medicine.

And yet its appeal is addictive. It is a human activity that is positive, active, aggressive and a wonderful mix of physical plus intellectual challenges which require clarity, imperative decision-making and equanimity during stress. Surgery of some sort appealed to me greatly.

At the MGH, Smith-Petersen made numerous remarkable contributions to the advancement of musculoskeletal care. Although rarely described as a clinician-scientist because the term did not exist during his era, he surely functioned as one. His contributions ranged from his eponymous anterior incision to the massive advance that arose from his development of the hip nail. The hip nail was quite remarkable because it led, for the first time in history, to the surgical management of both femoral neck fractures and intertrochanteric fractures, with huge advantages for the patients and for healthcare.

And finally he created the world's best intraarticular operation of its day for the reconstruction of severe osteoarthritis of the hip, namely the vitallium mold arthroplasty or "cup" arthroplasty. It is fascinating to realize that this whole concept grew from his own observations of the histology of the tissue reaction around an embedded piece of glass that he extracted from the back of a Harvard student on a Saturday afternoon in his Beacon Street office. His interpretation of that tissue response as a "synovial-like structure" producing joint-like fluid became the origin of his concept for a biologically regenerated hip joint.

Were these not enough contributions, the folklore in the Boston area would have you add what is often called the Darrach's operation, to his list of inventions. And also consider this. How well would you sleep the night before doing the world's first spinal ostomy, which he initiated?

At the time of my decision, the vast array of other compelling problems in musculoskeletal disease ranged from the steadily improving management of musculoskeletal tumors through the severe problems associated with infection, to demanding pediatric issues and intriguing problems of hand surgery, and on to the issues of fracture management. But, for me the challenges and opportunities of hip surgery were, by far, the most attractive and compelling. And besides, cup arthroplasty was not only a demanding operation with a prolonged, complex rehabilitation involving intense interpersonal relationships with each patient, but the results were variable. So there was a large opportunity for improvement.

Yet another factor in the particular profile which I elected to follow was the outstanding history at the MGH in major contributions to surgery in general and specifically to the understanding and management of hip disease. Dating back to the mid 19th century, the greatest singularity in surgical progress took place at the MGH nearly 170 years ago with the first public demonstration of the use of anesthesia. During that time, also, Henry Jacob Bigelow had generated innovative advances in the understanding of and management of traumatic dislocation of the adult hip. At the turn into the 20th Century E.A. Codman had (irritatingly) advanced critical, overpowering (but resisted) concepts into medicine, namely the urgent need for end result analyses and of registries of the outcomes of treatment. And in the 20th Century it was the work of Smith-Petersen which had been crucial in many aspects of musculoskeletal disease.

Thus it was that as I completed my training as Chief Resident in Orthopaedic Surgery at the MGH in 1959, two areas of major importance to the broad field of orthopaedic surgery attracted my focus. Clinically, the most demanding and rewarding region of specialization that presented an opportunity to me was that of reconstructive hip surgery.

But to decide where to aim the research efforts was virtually impossible. Still, I was guided by advice mentioned above that I have subsequently given unfailingly to sagacious young orthopaedic surgeons who ponder similar quandaries, "Go where no one else wants to go". Among orthopaedic surgeons at that time, that answer was metabolic bone disease. Osteoporosis was not only a massive riddle, it lacked both effective prevention and significant treatment. And yet, it contributed mightily to my commitment to solve problems involving the hip. Broadly speaking, research in quantifying the skeletal features of metabolic bone disease at that time consisted of two diverse movements, the vibrant and explosive new ability to quantify skeletal responses via radioactive isotopes and the innovative but more ponderous histologic approaches, typified by microradiography and other imaginative microscopic techniques.

Thus for the research portion of my focus, I chose within my initial commitment the area of the hip those worldwide, distressingly-common fractures which arose from age-related and postmenopausal osteoporosis.

Therefore I needed to obtain additional skills in research areas related to osteoporosis, specifically in metabolic bone disease with the special emphasis on bone formation, bone resorption, and the interrelationship between these two.

Since my senior literature thesis at Haverford College

summarized the worldwide use of radioactive isotopes in their role of providing astonishing new insights into physiology, I favored that route. However, entry into this field proved to be closed to a widely unknown, freshly-minted orthopaedic surgeon with this uncommon profile.

Plan B resulted in obtaining a National Infantile Paralysis Foundation grant for a fellowship, first at the Oak Ridge National Laboratory at Oak Ridge, Tennessee for an advanced course in nuclear physics, continuing on to the Royal National Orthopaedic Hospital, London. There the research environment focused on bone pathology, skeletal metabolism and the morphometric modes of quantifying skeletal responses.

My appointment there was in the Department with the wonderful name of Morbid Anatomy. It was under the direction of Herbert Sissons, an outstanding Australian musculoskeletal pathologist, and included Jennifer Jowsey, a D. Phil. from Oxford whose leadership in the quantification of formation rates of bone based on the new technique of microradiography was internationally recognized.

And yet, an overwhelming disappointment greeted my arrival in London at the Royal National Orthopaedic Hospital. They had no facilities for animal experiments. Even worse, they did not permit any research on any animals. My instinctive solution, namely simply to repair to the Royal College of Surgeons for such facilities, was equally thwarted by the identical limitation! In fact, the anti-vivisection sentiment was so strong throughout Britain that those limitations existed everywhere throughout the entire nation.

With one exception. That exception was Charles Darwin's home and farm. When Darwin died, he willed his home and experimental station to the nation under the one condition that experimental animal work be permitted there. And so it was that I found the single facility in all of Great Britain in which I could pursue experiments in dogs. It was located in Surry at Down, Darwin's home.

It was during that year that I was able to establish, from the work at Darwin's home, the valuable fluorescent microscopic method of quantification of rates of bone formation using multiple tetracycline labels, published in *Nature* in 1960.

Armed with these additional training and skills, after carefully considering other academic opportunities, in 1960 I returned to the MGH to join two world class movements of orthopaedic surgery. The first was the hip-centered, clinical office of Otto E. Aufranc, a key contributor to Smith-Petersen's career and subsequently his successor as the leading advocate of the cup arthroplasty in the world. And I became a member of the orthopaedic research staff in the freshly-minted basic science Orthopaedic Laboratory under the direction of Mel Glimcher. With remarkable brilliance as a scientist and an overwhelming personality, he had finished the clinical training in the Harvard Orthopaedics training program in Boston and crossed the Charles River to study molecular biology under Professor F. O. Schmitt at MIT. He returned to the MGH with these unique skills and catalyzed the development of an entirely new concept for basic orthopaedic research at the MGH. He also drove the effort for the funding of research space and the resources to allow this opportunity. This was achieved only through his inspired enthusiasm in conjunction with the relentless dedication, fund-raising and political skills of the Chief of Orthopaedic Surgery, Joseph Barr, the initiator of the concept of the ruptured intervertebral disc, and his right-hand deputy, Thornton Brown.

Glimcher rapidly assembled and organized a world-class unit, investigating with great success the basic nature of the specific and determinative role of collagen in its intimate dictate of that particular feature of the molecular structure of collagen which defined the location and nature of those calcium apatite deposits requisite to the creation of bone.

The times were electric, with incredible new insights into the basic nature and basic metabolism of the skeleton, and indeed many other aspects of the musculoskeletal system. It was enormously exciting to straddle both areas, namely the remarkable advances in the clinical management of musculoskeletal disease and the extraordinary innovations in the basic science of the musculoskeletal system.

Particularly challenging to me was yet another question, "what causes osteoarthritis of the hip". Why was it that so many patients wound up requiring major surgery after the progressive destruction of the cartilage in the hip joint. The overwhelmingly dominant contemporary theory held that it was the result of some ill-defined deficiency of the articular cartilage in the hip joint in that particular patient. But when assessed critically from the point of view of defining specifically that postulated cartilage abnormality, this theory was particularly unconvincing. Moreover, for me the link between the study of this problem and my compelling interest in the surgical reconstruction of the arthritic hip joint was a natural. While I loved all the clinical and surgical aspects of the care of patients with severe hip disease including the guidance and management of the patients and specifically the operative challenge of the reconstructive surgery, to me the question of the basic cause behind this disease was simultaneously uniquely compelling. At the Royal National Orthopaedic Hospital in London, Ronald Murray had already advanced his concept of the "tilt deformity" in the young adult male in relation to predicting who would develop osteoarthritis of the hip. If I could add to the understanding of why osteoarthritis of the hip would develop, perhaps I could reduce or even eliminate the very need for some, or even many cases of this reconstructive hip surgery.

It was into this stimulating environment at the MGH that the remarkable innovative surgical revolution called total hip replacement made its thrust. During surgery's most golden period, the seven decades from World War II until today, among the many awesome explosions of "unthinkable" surgical advances stands tall that singular, striking innovation which resulted in the "cure" of end stage arthritis of the hip, total hip replacement. It came to play a dominant role in this book.

## SELECTED RELATED REFERENCES

Mixter WJ, Barr J.S. Rupture of intervertebral disc with involvement of spinal canal. New Eng. J. Med. 211:210-215, 1934.

Smith-Petersen MN, Aufranc OE, Larson CB Principle of mold arthroplasty as applied to the hip. Surg Clin North Am. Oct 27:1303-1306, 1947.

Kuntscher G. Recent advances in the field of Medullary Nailing. Ann Chir et Gynaec Fenniae 37(2):115-136, 1948.

Barr JS. Low-back and sciatic pain: results of treatment. J. Bone & Joint Surg. 33A:633-649, 1951.

Glimcher MJ, Hodge AJ, Schmitt FO. Macromolecular Aggregation States in Relation to Mineralization: The Collagen-Hydroxyapatite System as Studied in Vitro. Proc Natl Acad Sci U S A Oct. 15;43(10):860-867, 1957.

Harris WH. A Microscopic Method of Determining Rates of Bone Growth. Nature 188: 1038-1039, 1960.

Smith-Petersen MN, Larson CB, Aufranc OE. Osteotomy of the spine for correction of flexion deformity in rheumatoid arthritis. Clin Orthop Relat Res. Sept-Oct; 66:6-9, 1969.

# CHAPTER 1

# SKELETAL RENEWAL AND METABOLIC BONE DISEASE

THE FUNDAMENTAL IMPORTANCE OF SKELETAL metabolism as the basis underlying the extremely common osteoporotic fractures in the elderly was well projected in our medical progress report "Skeletal Renewal and Metabolic Bone Disease" in the New England Journal of Medicine five decades ago:

"The skeleton, containing 99% of the total volume of calcium, serves two major functions. First of all, it plays an important role in calcium homeostasis, both responding to and contributing to changes in calcium metabolism. Secondly, the structural integrity of the skeleton is essential for normal existence. Fractures, by far the most important abnormality of the skeletal system, occur with increasing frequency in the elderly because of decreasing strength of the skeleton. This weakness is due largely to a reduction in skeletal mass caused by an imbalance between formation and resorption of bone. Throughout life, even after cessation of longitudinal growth, cancellous and cortical bone are constantly being replaced by resorption of existing areas and production of new deposits in microscopic amounts at many sites heterogeneously distributed throughout the shell. Changes in this balance between formation and resorption have a critical role in calcium homeostasis and underlie every disease with a notable influence on the adult skeleton."

Age-related and postmenopausal osteoporosis are dominant contributors to fractures of the spine, hips, wrists, shoulders, and many other areas in the elderly. The incidence of hip fractures and vertebral fractures in the elderly group remains high in the United States and around the world even today, despite the availability of resorption blockers. It is estimated that only 25 % of women needing these medications are taking them, currently.

Because of this massive importance both to patients and to the management of musculoskeletal disease and trauma worldwide, studies of skeletal renewal became a major thrust of my research beginning in 1959.

The two major methods of studying the rates of skeletal renewal in vivo were calcium-kinetic methods and morphometric methods. While the calcium-kinetic methods arising de novo from the relatively new (at that time) techniques of radioactive isotope studies had been elements of my senior thesis for Honors in Chemistry as an undergraduate at Haverford College, I focused instead on our unique morphometric method of quantifying localized rates of new bone formation. This quantification was made possible by adding the innovation created by the detection of focal, microscopic, detectable images of every site of new bone formation by the uptake of tetracycline molecules, revealed by fluorescence microscopy.

This technique allowed the direct quantification of *rates* of bone formation locally in various parts of the skeleton under diverse conditions. However, basic to the realistic quantification of these important rates was the critical determination of how much variation occurred at a given site over time as well as how much variation occurred from site to site in the same animal at the same time. I studied this issue intensively.

When these sets of studies and others were combined with our detailed calcium-kinetic studies in dogs done at the MGH in conjunction with Robert Heaney from the Department of Medicine of the School of Medicine, Creighton University, Nebraska and melded with Heaney's clinical and research studies of human skeletal metabolism, the results stimulated our three-part Medical Progress Report in the New England Journal of Medicine. This set of articles defined the current state of skeletal metabolism for that era and introduced the compelling thesis of the highly interrelated connection between formation and resorption over a wide variety of conditions in the overall assessment of skeletal remodeling.

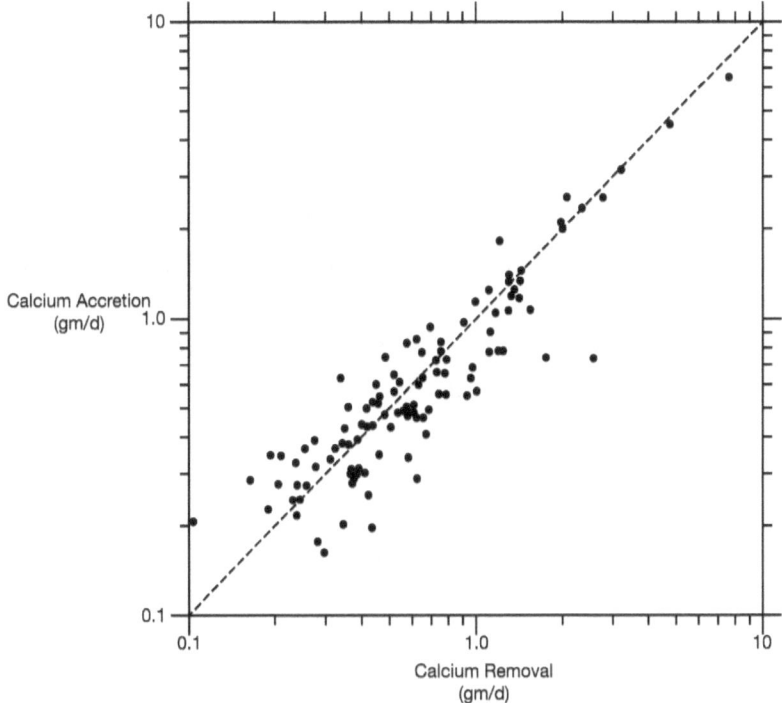

**Figure 1.1. Calcium Removal from bone plotted as a function of Calcium Accretion in 108 patients with wide variety of orders affecting calcium metabolism.** From Harris, W.H., Heaney, R.P.: Skeletal Renewal and Metabolic Bone Disease. A Medical Progress Report. N Engl.J.Med., 280:204. Copyright 1969 Massachusetts Medical Society. Reprinted with permission.

Our studies of calcium kinetics in dogs reinforced the concepts of the close coupling of resorption and formation proposed in this article and our growth hormone studies in dogs demonstrated growth hormone to be the only known agent at that time, with the exception of toxic doses of fluorides, that would directly stimulate net new bone formation.

Subsequently, through an entirely different mechanism, other investigations demonstrated that it was possible to effect a major reduction of the skeletal bone loss with increasing age in adult women and, importantly, substantial reduction in the incidence of fractures secondary to age-related osteoporosis. This was done by the striking development of the inhibition of resorption in humans by bisphosphonates, drastically changing the risk of fracture. This approach prompted far-ranging developments in these types of compounds while leading to, in many series, a decrease in fracture rates among osteoporotic patients of 30 to 50%, including vertebral fractures, hip fractures and others. While these compounds are not without specific adverse side effects, this was such a dramatic result at that time that I then made the decision to shift my focus away from the field of skeletal metabolic studies.

But before completing that shift, I had demonstrated the interesting findings on the ability of alizarin red S and selected tetracycline compounds to effectively partially inhibit the initial deposition of calcium at sites of new bone formation.

In addition I identified in our growth hormone studies the unique dissociation between the effect of exogenous extra growth hormone in intact adult dogs on the differential response of articular cartilage compared to that of the skeleton. The the skeleton showed a marked increase in new bone formation from exogenous growth hormone while detailed studies of the articular cartilage showed no response, suggesting that the osteoarthritic changes in acromegaly were not related to a direct stimulus to the articular cartilage by the increased levels of growth hormone.

We also used this microscopic morphologic technique of quantifying rates of bone formation to investigate another fracture-related problem, specifically, quantifying the effect of compression alone on new bone formation during compression plating in the intact canine femur, independent of the confusing influence of fracture healing. The simple act of applying a plate and screws to the intact tibia substantially stimulated new bone formation, but by 12 weeks any increase in resorption had not become statistically significant. Surprisingly, the addition of <u>compression</u> <u>itself</u> nor of distraction did not influence these changes. This increase in new bone formation was caused by simply the application of the plate and screws, not by compression.

## SELECTED RELATED REFERENCES

Harris WH. A Microscopic Method of Determining Rates of Bone Growth. Nature 188: 1038-1039, 1960.

Harris WH, Jackson RH, Jowsey J. The In Vivo Distribution of Tetracyclines in Canine Bone. J. Bone Joint Surg., 44-A: 1308-1320, 1962.

Jacobs R, Harris WH, Katz EP, Glimcher MJ. The Interaction Between Tetracycline and Reconstituted Guinea-Pig-Skin Collagen In Vitro. Biochim Biophys. Acta 86:579-587, 1964.

Harris WH, Travis DF, Friberg U, Radin E: The In Vivo Inhibition of Bone Formation by Alizarin Red S. J. Bone Joint Surg., 46-A: 493-508, 1964.

Harris WH, Nagant de Deuxchaisnes C. Controlled in Vivo Inhibition of Bone Formation. In: Proceedings of the Second European Symposium on Calcified Tissues, University of Liege, Ed. LJ Richelle and MJ Dallegmagne, 93-197, 1965.

Harris WH, Haywood EA, Lavoria J, Hamlen DL. Spatial and Temporal Variation in Cortical Bone Formation in Dogs. J Bone Joint Surg., 50-A:1118-1128, 1968.

Harris WH, Haywood EA, Hamblen DL, Lavorgna J. Analysis of Covariance of Bone Formation Rates by Specific Skeletal Sites and Over Time. In: Proceedings of the Fifth European Symposium on Calcified Tissues, Bordeaux, Societe d'Edition d'Enseignement Superieur, Editors: G. Milhaud, M.Owen, J.J. Blackwood:415-421, 1968.

Harris WH, Lavorgna J, Hamblen DL, Haywood EA. The Inhibition of Ossification in Vivo. Clin. Orthop. Rel Res, 61:52-60, 1968.

Harris WH, Heaney RP: Skeletal Renewal and Metabolic Bone

Disease. A Medical Progress Report. N Engl. J. Med., 280:193-202, 253-259, 303-311, 1969.

Galante J, Rostoker W, Lueck R, Ray RD. Sintered fiber metal composites as a basis for attachment of implants to bone. J Bone Joint Surg, Jan ;53(1):101-114, 1971.

Harris WH, Heaney RP, Jowsey J, Cockin J, Akins C, Graham J, Weinberg EH. Growth Hormone: The Effect on Skeletal Renewal in the Adult Dog. Part I. Morphometric Studies. Calcif. Tiss. Res., 10:1-13, 1972.

Heaney RP, Harris WH, Cockin J, Weinberg EH: Growth Hormone. The Effect on Skeletal Renewal in the Adult Dog. Part II. Mineral Kinetic Studies. Calcif. Tiss. Res., 10:14-22, 1972.

Coutts RD, Harris WH, Weinberg EH: Compression Plating: Experimental Study of the Effect on Bone Formation Rates. Acta Orthop. Scand.,44:256-262, 1973.

Harris WH, Heaney RP, Weinberg EH, Cockin J, Akins CM, Graham J. Skeletal Responses of Young Adult and Aged Dogs to Growth Hormone. In: Proceedings of the 9th European Symposium on Calcified Tissues. Vienna: Verlag H. Egermann Facta-Publication, Ed. H Czitober, J Eschberger. 75-80, 1973.

Mankin, H.J., Zarins-Thrasher, A., Weinberg, E.H., Harris, W.H.: Dissociation between the Effect of Bovine Growth Hormone in Articular Cartilage and in Bone of the Adult Dog. J Bone Joint Surg., 60-A: 1071-1075, 1978.

Fleisch H. Development of Bisphosphonates. Breast Cancer Res 4(1) 30-34, 2002.

Eriksen EF, Diez-Perez A, Boonen S. Update on long-term treatment with bisphosphonates for postmenopausal osteoporosis: a systematic review. Bone 58:126-135, 2014.

Adler R, Fuleihan E-H, Baue D, Camacho P, Clarke B, Clines G, Gregory A, Compston J, Matthew T, Edwards B, Favus M, Murray J, Greenspan S, McKinney R, Pignolo R, Sellmeyer D. Managing Osteoporosis in Patients on Long-term Bisphosphonate Treatment: Report of a Task Force of the American Society for Bone and Mineral Research. Journal of Bone and Mineral Research 31(1): 16-35, 2016.

# CHAPTER 2

---

# THE ETIOLOGY OF OSTEOARTHRITIS

WHY DOES THE HIP JOINT wear out? What is the etiology of osteoarthritis of the hip?

Fifty-seven years ago, in 1960, as I contemplated the opportunities for an academic career that lay ahead of me, the prospects were attractive. I was working at a premier teaching hospital, the MGH, working clinically on hip reconstructions in Otto Aufranc's world-famous orthopaedic office founded by Marius Smith-Petersen, and working experimentally in the Orthopaedic Research Lab at the MGH. At least, so it seemed, until that reverie was disturbed by an errant thought. Why is it that you wait until the hip joint is destroyed and then do major surgery to repair that damage? Why not prevent the arthritis? But to be able to do that, you need to know what actually causes the problem.

I had witnessed first-hand the extraordinary and dramatic cure of an entire disease, poliomyelitis. This remarkable transition had occurred immediately following my introduction to the Harvard Orthopaedic Training Program in 1955 at the Children's Hospital of Boston. 1955 seared the memory of all of us as the year of a massive polio epidemic. Although I had been originally assigned to a research position, the epidemic – which closed all Boston hospitals to the treatment of anything except polio and emergencies – demanded that, like everyone else, I participate in the care for those <u>1800</u> children admitted in just one month (August) to the Children's Hospital with polio, averaging 60 fresh cases per day. Eighteen hours a day, all of us treated polio patients. It was a nightmare for the patients, families, and the town.

The following year there were no cases! The vaccines had taken effect.

The partial restoration of function in the limbs and trunk of polio patients had been the lifeblood of pediatric orthopaedic surgery. The opportunities and the results of muscle transfer and specialized fusions were gratifying to the patients and rewarding to the surgeons. But what an astonishing example was the difference between 1955 and 1956! How stunning was the difference between <u>prevention</u> of polio versus <u>treatment</u> of polio. Because I experienced first-hand this extraordinary transition in the front lines of that change, its impact on me was powerful and long-lasting.

Now – five years later as I began my own career – the obvious observation that it is far better to <u>prevent</u> than <u>treat</u> disease – remained compelling and rang in my ear.

This specific question about the cause of osteoarthritis of the hip had been stimulated in my thinking by a remarkable radiologist, Ronald Murray, during my postdoctoral year at the Royal National Orthopaedic Hospital in London. He advanced the insightful idea that by analyzing the x-ray of the hip in a young adult, he could predict who would develop osteoarthritis of the hip many years later. Astonishing. He had identified certain abnormal appearances of the hip joint, specifically the contours of the proximal portion of the femur – which he designated the "Tilt Deformity" – occurring in otherwise "normal" hips in fully active, vigorous healthy young adult males. This radiographic sign, the "Tilt Deformity", had powerful, accurate predictive value for late onset adult of osteoarthritis of that joint (1).

What, in fact, he had observed was that those young adults (mostly males) had the radiographic contours of the proximal femur which reflected that the patient had probably had a minor but clinically silent slipped capital femoral epiphysis during adolescence. It was this pathology – this abnormality in the development of the proximal femur – which condemned them to the arthritic destruction of the hip joint many years later.

For me, that he could identify a predictive factor – an x-ray appearance in this case – but any predictive factor that indicated reasonably accurately the later onset of osteoarthritis of the hip, was miraculous. Could his experience be confirmed? Why, how, and what other indicators might exist?

As my clinical practice grew, I pursued Murray's thought about a predictive radiographic appearance of not only the "Tilt Deformity" but also as an extended search for any other predictive configurations. To do this I initiated a rigorous program to recover any and all prior x-rays from any source or site on all the adult patients who presented with arthritis of the hip. Not surprisingly, this effort paid off in two ways. I strongly confirmed Murray's suggestion plus I began to identify many other recognizable, slightly abnormal features which were distinctly different from the tilt deformity but also had predictive power.

Prominent in the group were subtle suggestions of evidence of developmental dysplasia, so subtle that, without a specific interest in such findings, it was easy to pass these changes as being within normal limits (2).

When these patients were then analyzed as a group, other key characteristics became obvious. In contrast with those who demonstrated the tilt deformity, who were predominantly male and often unilateral, these latter patients were mainly female and commonly had bilateral abnormalities.

Then additional other subtle manifestations of different developmental abnormalities became identified, again often overlooked as being normal. They included subtle changes caused by Legg-Perthes disease, multiple epiphyseal dysplasia, spondyloepiphyseal dysplasia, early avascular necrosis of the femoral head and the intra-acetabular labrum (3). And in addition to gender differences, other important factors such as age of onset of the arthritis became differentiating features.

As this data base grew, it then became clear that if one eliminated extraneous features such as joint infection and major hip trauma, as well as deleting patients with inflammatory conditions such as rheumatoid arthritis or lupus and very unusual diseases like ochronosis and hemochromatosis, virtually all adult patients presenting with osteoarthritis of the hip had a distinct, although often subtle, abnormality in development of the joint. (3) (4) (5) (6). In other words, there were virtually none with a truly normal hip initially. This had huge implications. If confirmed, it could mean that there was no such disease as "primary osteoarthritis" of the hip. This would mean rejection of the worldwide belief that osteoarthritis of the hip resulted from some poorly described abnormality of the cartilage of the hip, a cartilage disease. If true, it also raised the very exciting possibility that correction of whatever deformity existed – could delay or prevent the development of the arthritis. What a prospect!

For centuries physicians knew that the severe forms congenital or developmental abnormalities of the hip presaged the onset of osteoarthritis later in life and even at times the early onset of osteoarthritis. What we were now recognizing was different, very different, and critically important. This new idea postulated that even minor, very subtle abnormalities would lead to the same result, although perhaps decades later. Virtually all adult patients with true osteoarthritis of the hip had some degree of deformation of the sphericity of the hip joint, femoral head and neck or acetabulum, or a combination deformity. Arthritis developed early with severe deformities and much later with subtle deformities. An underlying abnormality of the cartilage per se did not appear to be a factor.

Louis Solomon, Chief of Orthopaedics at the University of Witwatersrand in South Africa, also having been influenced by Murray, conducted similar studies at Johannesburg. His findings were not only confirmatory, he also found out that the racial differences in the incidence of the developmental disease types of the hip were reflected in a parallel frequency in adult osteoarthritis of the hip, strongly supporting this radical concept (7-11).

To advance such a huge change in belief about the etiology of osteoarthritis of the hip prompted dispute and disbelief, particularly among arthritis experts and cartilage scientists. In nearly all series of cases of osteoarthritis of the hip at those times, the leading diagnostic category was "primary osteoarthritis", in direct contradiction to this proposed concept which actually proposed denying that such a condition existed (4) (5).

Because our observations were retrospective, i.e., made by analysing the early radiographics of patients presenting with osteoarthritis of the hip late in life, we sought prospective prediction for confirmation. For that we allied with the Alfred I. duPont Hospital for Children in Wilmington, Delaware, because of their excellent preservation of early x-ray films of patients with hip disease coupled with their outstanding long-term follow-up radiographic data. From the series of studies arising from this resource, we generated confirmation of our hypothesis (2) (6).

While the acceptance of this concept grew slowly but progressively, the lack of a practical, surgical mode of correcting the abnormalities, an especially challenging proposition in young asymptomatic patients, remained a major obstacle. This deficiency was rapidly overcome by the imaginative work of Ganz and followers (12) (13) who developed both the safe technique of surgical dislocation of the hip and the concept of "femoral-acetabular impingement" (FAI) as the mechanism by which these subtle developmental abnormalities of the hip produced the arthritic change, independent of the mode which produced the abnormality.

Today, the acceptance of this concept of the etiology of osteoarthritis of the hip is worldwide and the operative management of FAI is, putatively, the second most important development in comprehension and management of osteoarthritis of the hip of the 20th century, second only to the enormous revolution resulting from the invention and development of total hip surgery. Surgical correction of FAI has produced extensive relief of symptoms associated with these developmental abnormalities and will with further time and follow-up data likely be able to establish a quantifiable reduction in the progression of some of these abnormalities to produce osteoarthritis.

## SELECTED RELATED REFERENCES

1.  Murray RO. The etiology of primary osteoarthritis of the hip. Br J Radiol. 35:810-824, 1965.
2.  Stulberg SD, Harris WH. Acetabular Dysplasia and Development of Osteoarthritis of the Hip. In: The Hip, Proceedings of the Second Open Scientific Session of the Hip Society, Ed. W. H. Harris, St. Louis: C.V. Mosby Co., pp. 82-93, 1974.
3.  Harris WH, Bourne RB, Oh I. Intra-articular acetabular labrum: a possible etiological factor in certain cases of osteoarthritis of the hip. J Bone Joint Surg. Am;61:510-514, 1979.
4.  Harris WH. Primary osteoarthritis of the hip: a vanishing diagnosis. J Rheumatol. Suppl 9:64, 1983.
5.  Harris WH. Etiology of osteoarthritis of the hip. Clin Orthop Rel Res. 213;20-33, 1986.
6.  Stulberg SD, Cordell LD, Harris WH, Ramsey PL, MacEwen GD. Unrecognized Childhood Hip Disease: A Major Cause of Idiopathic Osteoarthritis of the Hip. In: The Hip, Proceedings of the Third Open Scientific Meeting of the Hip Society, Ed. HC Amstutz, St. Louis: C.V. Mosby Co., pp. 212-228, 1975.
7.  Solomon L, Beighton P. Osteoarthritis of the hip and its relationship to pre-existing in an African population. J Bone Joint Surg Br. 55:216-217, 1973.
8.  Solomon L. Patterns of osteoarthritis of the hip. J Bone Joint Surg Br. 58:176-183, 1976.
9.  Solomon L. Studies on the pathogenesis of osteoarthritis of the hip. Trans Coll Med South Africa:104-124, 1981.

10. Solomon L, Schnitzler CM, Browett JP. Osteoarthritis of the hip: the patient behind the disease. Ann Rheum Dis. 41:118-125, 1982.1. Solomon L. Geographical and anatomical patterns of osteoarthritis. Br J Rheumatol;23:177-180, 1984.

11. Ganz R, Gill TJ, Gautier G, Ganz K, Krugel N, Berlemann U. Surgical dislocation of the adult hip: A technique with full access to the femoral head and acetabulum without the risk of avascular necrosis. J. Bone Joint Surg. BR 83:1119-1124, 2001.

12. Ganz R, Leunig M, Leunig-Ganz K, Harris WH. The etiology of osteoarthritis of the hip: an integrated mechanical concept. Clin Orthop. Rel Res. 466(2):264-72, Feb 2008.

# PHYSICAL MICROENVIRONMENT WITHIN HUMAN ACETABULAR ARTICULAR CARTILAGE IN VIVO

A BETTER UNDERSTANDING OF THE physical microenvironment within human articular cartilage is central to both the physiological functioning of human articular cartilage and the pathology leading to and resulting from osteoarthritis of the hip. The wide range of human hip positions, functions, and loads adds greatly to the range of conditions imposed on this cartilage, as well as the direction and location of the loads applied. All of these factors are greatly multiplied by any errors in growth, development or congruence of the hip joint and, of course, become compounded by the progressive deterioration of the cartilage as arthritis develops.

In the past, considerations of the physical microenvironment of human acetabular articular cartilage have been predominantly based on localized static tests or tests of extracted samples of cartilage. Others were conceptual in nature. Peculiarly absent in these considerations are the missing but important muscular co-contractions occurring across the joint, because they cannot be quantified accurately using the customary inverse Newtonian analysis. A fresh approach would permit new insights into these important factors.

To address all these important, unresolved considerations,

Professor Robert Mann of Mechanical Engineering at MIT attacked this knowledge deficit by envisioning a unique precision measuring device in the form of a femoral endoprosthesis with multiple pressure gauges built into the sphere, emitting readings from an encapsulated, internal sampling radio telemetry system within the ball of the endoprosthesis, in short, the creation of an instrumented endoprosthesis. To these data he also added the concept of detailed quantification of the thickness of the entire, global intact acetabular articular cartilage, both for geometric purposes and to study the influences of the subchondral bone on the local pressure relationships.

Once inserted into the hip of an elderly patient requiring an endoprosthesis for the surgical management of a displaced femoral neck fracture, direct quantifications of the pressure on the intact human acetabular articular cartilage during and after recovery from the surgery could be recorded during the diverse activities of daily living.

After we selected a chrome-cobalt endoprosthesis as the vehicle, the early work centered on creating the pressure sensors in the wall of the ball. Since these prostheses are constructed in two halves which subsequently are screwed together and then welded shut, all the important sites for the pressure sensors could be located in the upper half of the sphere.

The concept was to reduce the thickness of the wall of the chrome cobalt sphere at each desired site to a degree of thickness such that the remaining wall of the sphere at that site would be sufficiently responsive to the external forces that its displacement could be translated into an electrical signal via a push rod at right angle to a strain gauge transducer attached to the inner surface of the wall. The critical determination was that the fatigue strength of this thin detection site would be adequate after it had been reduced in thickness sufficiently to be sensitive to the range of pressures it would experience during the activities of daily living.

Fortunately, the chrome cobalt material used in the spheres of these endoprostheses met the required fatigue strength during extensive testing, even after being reduced in thickness to the point that allowed appropriate deflection enough to activate the strain gauges proportionally under our estimated range of pressures likely to be encountered.

Fourteen distributed sites for pressure sensors on the sphere were selected. In retrospect, an important deficit occurred in that no arrangement was made to be able to integrate the data from these fourteen sites into a single resultant force across the hip joint. However, in all other respects the electronics were remarkable, designed to sample each site 250 times a second. All the electronics were contained within the hollow sphere of the femoral head of the prosthesis.

Two special problems were <u>sterilization</u> of all these electronics in the sphere to eliminate the risk of infection from the components of the prosthesis and establishing <u>power</u> for the unit. A power induction scheme solved the power situation, eliminating the need for either batteries or for direct wire connections through the skin as some other investigators had chosen. Using a power induction scheme had great advantages, with the stem of the prosthesis serving as the antenna for both power induction and for broadcasting the data back out. The requirements for these techniques substantially modified the design of the electronics, but this approach avoided the risk of infection being introduced along wires penetrating the skin and also avoided the limitations of built-in batteries.

The sterilization was achieved by leaving an access port to the inferior portion of the sphere after all the pressure sensors had been created. Sterilizing gas was then introduced to the interior of the sphere through the access port to sterilize the interior of the sphere and the electronics, so that, should anything adverse happen that exposed joint fluid to the electronics, infection would be unlikely to occur. All of this was done in a Class I environment. Once the sterilization had occurred, the access hole was plugged with a chrome cobalt bung, which was then sealed into the chrome cobalt sphere. After the electronics had been placed in the hollow sphere and after the two halves of the sphere had been reassembled, the sphere was welded shut and sterilized. The last step required special care and techniques so that the welding did not disturb the pressure sensors or the electronics.

Two important and obvious limitations existed to this concept of measurement of the pressures in the articular cartilage of the hips in vivo. The first was the construction of only a single size of prosthesis, because of its cost and complexity. This meant that the insertion of the prosthesis required a patient who needed exactly this size and no other. The second requirement was for a patient who comprehended the importance of this project to be willing to volunteer to have this unique, experimental prosthesis inserted into his or her body instead of a standard endoprosthesis. Some advantage in approaching this problem was presented by knowing the profile of the patients most likely to need an endoprosthesis, namely elderly females, i.e., those who most commonly sustain a femoral neck fracture. And from the endoprostheses manufacturers we learned the specific size of the most often inserted endoprostheses. Still, to find an elderly female with a femoral neck fracture who had the experimental spirit deeply embedded enough to entertain the acceptance of a "never used before" prosthesis and to agree to participate in an ongoing, lifetime experiment requiring repeated visits to the locomotion lab for detailed, intense gait studies, remained a continuing concern throughout the long developmental phase of the work.

Moreover, detailed exploratory studies showed that the thickness of the cartilage layer in the human intact acetabulum was not uniform and that the global profile of the abrupt transition from the cartilage to the rigid, mineralized subchondral condensation was characterized by substantial peaks and valleys on the microscopic scale. These rather acute changes in thickness and mineralized subchondral condensation altered the local pressure relationships sharply and substantially. That led to imaginative, detailed studies of these variables via the development of ultrasound methods to quantify the thickness of the cartilage at individual sites on the acetabulum and simultaneously characterize the profile of the subchondral bone.

These ultrasound data demonstrated that while the cartilaginous surface of the acetabulum was spherical to within 150 microns, that was not so for the underlying calcified surface, which varied as much as 500 microns in height. Moreover, these two global contours were actually slightly offset from each other. These studies allowed this information to be correlated with the pressure readings, enhancing the validity of the changes in pressure readings and explaining the source of many of the variations.

Prior to our findings, the two primary theories of pressure distribution across the human acetabular cartilage were a) that an axisymmetric high peak pressure of unknown magnitude existed at the center of the expanse of cartilage in contact with the femoral head and that this pressure progressively diminished to zero at the edge of contact with the femoral head articular cartilage, and b) that the opposing speculation conferred properties on articular cartilage which effected a relatively low and relatively uniform distribution of pressures throughout the contact surfaces which dissipated gradually at the edges of contact. Neither of these hypotheses addressed specifically changes in pressure and pressure distribution with different human activities or positions, nor could they account for the effect of muscular contractions across the hip joint.

Another contributor to the long duration of this project was the simultaneous development of extensive advances in the study of human gait by Professor Mann. While force-plate techniques had existed and tracking programs had proliferated for years, he advanced this arena substantially by developing two key capacities, the ability to record and project both limbs during the walking cycle and the techniques of quantifying the third dimension of the limbs, thus adding quantification of abduction-adduction and of hip rotation to the simpler stick figures commonly used to illustrate the results of such gait studies. This major enhancement of gait studies contributed substantially to understanding and coordination of the pressure measurement during the activities involved and the position of the limb relative to the hip joint.

Using an uncomplicated computer-driven hip simulator containing a simplified instrumented endoprosthesis, studies were done on autopsy-retrieved acetabula. These produced the important results, in addition to the ultrasound data, about the consolidation of human intact articular cartilage under spherical load and also critical information about the need for accurate sizing of endoprostheses by hip surgeons in their clinical practice. These latter studies showed that the pressures in the acetabular articular cartilage doubled if the prosthesis was mismatched to the acetabular size by just 2 mm. Since, in those days, endoprostheses were manufactured only in 2mm increments, these observations contributed to a better understanding of the frequent penetration of endoprostheses into and through the acetabular cartilage during clinical use, leading to clinical failures as was commonly seen as a complication of the insertion of endoprostheses.

From these data came multiple recommendations to surgeons for a variety of better techniques to improve the accuracy of deciding the optimum size of endoprostheses to use during insertion of an endoprosthesis for treatment of a displaced femoral neck fracture. These same techniques were used to quantify the determination of the size that our selected patient required to assure that she did have the optimum acetabular size to receive this instrumented prosthesis. Moreover, on the basis of these determinations orthopaedic manufacturers doubled the increment of sizes available to the surgeons to 1 mm steps.

The studies from the patient who entered the experiment, a 73-year-old female who had sustained a displaced femoral neck fracture, were striking. They provided unique information about the physical microenvironment of the intact human acetabular articular cartilage in vivo under a wide variety of activities, using integration of the pressure data with the concurrent kinetic-kinematic data. Pressure studies were done on her over three years, especially in order to quantify the changes occurring throughout the full period of rehabilitation. Correlation of the pressure data with the kinematic data including the foot-floor force vector data produced unique results. Critical data on co-contraction were generated.

The peak pressure recorded was a remarkable 18 MPa and was directed posteriorly, since this peak value occurred during rising from a standard chair. Rising from a chair produced peak pressures three times that of during walking and even pressures which were higher than those associated with either jogging or jumping. Climbing stairs also produced high pressures, as much as 10 MPa.

And throughout all of these data a most consistent finding was the great local variability of the pressure readings because of the variations in cartilage thickness and peaks in the subchondral mineralized layer.

These studies refuted prior suggestions that the pressure on intact human acetabular articular cartilage was relatively uniform or axisymmetric. They are highly irregular, with pressures ridges and valleys. The maximum pressures recorded vastly exceeded those previously estimated. Also the shapes of the pressure distribution strongly correlated with the shapes of the cartilage compression distribution.

Even in the early rehabilitation period such activities as using a bedpan produced high cartilage pressures, equal to those found during walking even after full rehabilitation.

Longstanding questions of rehabilitation were answered. While strong muscular forces are required to walk non-weight bearing on two crutches, they do not overcome the advantage of the crutches, which result in a low pressure during ambulation, reducing the peak pressure in half. Partial weight bearing on two crutches decreases the peak load by one-third. A cane in the opposite hand, even with maximum force on it, only reduces peak load minimally, by about 10%.

Other features of the early rehabilitation period of interest were the quantification that balanced suspension did reduce peak pressures lower than bed rest alone and that even at bed rest, isometric contractions could produce pressure nearly equal to those found during gait.

In terms of important changes in the practice of hip surgery, in addition to the valuable other new insights concerning anatomy, physiology and functioning of human cartilage that followed from these data, were the recognition of the importance of and the methods of accurately determining sizing of endoprostheses of the hip, the change in sizes available to the surgeon, and changes needed in supporting the posterior aspect of the acetabulum in complex reconstructions involving deficiencies in that posterior aspect.

The realization that human intact acetabular cartilage sees peak pressure as high as 18 MPa and commonly up to 10 MPa occur routinely in stair climbing are among the most important findings.

While the design of the instrumented prostheses did not permit direct quantification of the magnitude of the cocontractions, irrefutable information as to their nature was obtained. The pressure magnitude measured clearly exceeded the data calculated from the inverse Newtonian calculations and major differences existed in the location and directions of the peak pressures measured compared to those calculated from external data. Moreover during gait, hip pressures rose <u>prior</u> to foot-floor contact, specifically just prior to the end of swing phase, and descending stairs but not prior to ascending stairs.

As an aside, this project reflected certain unspoken truths about research. When the complexities of the experimental design (building a Buck Rogers hip prosthesis that would broadcast from within the body the human articular cartilage pressure data in vivo during multiple activities over several years) were coupled with its expense and the severe difficulties in obtaining funding, the project required 17 years to complete. Hidden in that startling figure are two observations. First, it is sometimes better <u>not</u> to know in advance how long a project will take, for otherwise it is likely to be abandoned. Secondly, this illustrated clearly the importance of commitment and persistence to bring such a research project to successful completion.

## SELECTED RELATED REFERENCES

1. Harris WH. Sinking Prostheses. Surg. Gynecol. Obstet., 123: 1297-1302, 1966.
2. Carlson CE, Mann RW, Harris WH. A Look at the Prosthesis-Cartilage Interface: Design of the Hip Prosthesis

Containing Pressure Transducers. J. Biomed. Mater. Res. 8:261-269, 1974.

3. Carlson CE, Mann RW, Harris WH. An Experimental Technique for Measuring the Cartilage Surface Pressure Distributions in the Hip. New England Conference on Bioengineering. Ed. M.H. Pope, RW McLoy, and RG Abshaer, University of Vermont, pp. 90-96, 1974.

4. Carlson EC, Mann RW, Harris WH. A Radio Telemetry Device for Monitoring Cartilage Surface Pressure in the Human Hip. Biomed. Eng. BME-21:257-264, 1974.

5. Harris WH, Rushfeldt PD, Carlson CE, Scholler J-M, Mann RW. Pressure Distribution in the Hip and Selection of Hemiarthroplasty. In: The Hip, Proceedings of the Third Open Scientific Meeting of the Hip Society. Ed: H.C. Amstutz, St. Louis: C.V. Mosby Co., pp.93-98, 1975.

6. Harris WH, Rushfeldt PD, Carlson CE, Scholler J-M, Mann RW. Pressure Distribution in the Hip and Selection of Hemiarthroplasty. In: The Hip, Proceedings of the Third Open Scientific Meeting of the Hip Society. Ed: HC Amstutz, St. Louis: C.V. Mosby Co., pp.93-98, 1975.

7. Carlson CE, Mann RW, Harris WH. Design Features of a Pressure Telemetering Hip Prosthesis. Biotelemetry, III 127-130, 1976.

8. Oh I, Carlson CE, Tomford WW, Harris W. Improved Fixation of the Femoral Component after Total Hip Replacement Using a Methacrylate Intermedullary Plug. J. Bone Joint Surg., 60-A: 608-612, 1978.

9. Mann R, Rushfeldt P, Harris WH. Influence of Cartilage Geometry on the Pressure Distribution in the Human Hip Joint. Science, 204:413-415, 1979.

10. Rushfeldt PD, Mann RW, Harris WH. Improved Techniques for Measuring In Vitro the Geometry and Pressure Distribution in the Human Acetabulum I. Ultrasonic Measurement of Acetabular Surfaces, Sphericity, and Cartilage Thickness. J. Biomechanics, 14:253-260, 1981.

11. Rushfeldt PD, Mann RW, Harris WH. Improved Techniques for Measuring in Vitro the Geometry and Pressure Distribution in the Human Acetabulum II. Instrumented Endoprosthesis Measurement of Articular Surface Pressure Distribution. J. Biomechanics, 14:315-323, 1981.

12. Hodge WA, Carlson KL, Fijan RS, Burgess RG, Riley PO, Harris WH, Mann RW. Contact pressures from an instrumented hip endoprosthesis. J. Bone & Joint Surg. Am. 71-A:1378-1386, 1989.

13. Krebs DE, Robbins CE, Lavine L, Mann RW. Hip biomechanics during gait. J Orthop Sports Phys Ther. 28(1):51-59, 1998.

14. Park S, Krebs DE, Mann RW. Hip muscle co-contraction: evidence from concurrent in vivo pressure measurement and force estimation. Gait and Posture 10: 211-222, 1999.

15. Morrell, K, Hodge WA, Krebs DE, Mann RW. Corroboration of in vivo cartilage pressures with implications for synovial joint tribology and osteoarthritis causation. Proc Natl Acad Sci USA 102(41):14814-14824, 2005.

# CHAPTER 4

‒ ‒ ‒ ‒ ‒ ‒ ‒ ‒ ‒ ‒ ‒ ‒ ‒ ‒ ‒ ‒ ‒

# REPLANTATION OF HUMAN LIMBS

MAY 23ʳᵈ 1962 A 12-YEAR-OLD boy attempted to hitch a ride on a passing freight train. His right arm was amputated between the side of the train and a stone abutment, at the level of the deltoid insertion. He sustained no other injuries except to the tips of his left thumb, index and middle fingers.

Two unexpected events occurred near that time, one preceding his injury and one following it. Earlier that year a surgeon at the MGH, R.S. Shaw, had addressed Orthopaedic Grand Rounds proposing that since all the necessary techniques existed currently to repair severed vessels, bones, muscles, nerves, skin and soft tissue and since successful repairs were being done to very extensive injuries to extremities that were close to but not quite amputated, it should be possible to replant a completely severed limb. The other paraaccidental curious event was that the rescue personnel attending this boy elected to send his amputated arm along with the patient to MGH, an unlikely decision since replantation of an amputated limb had never been done. Their reasoning was never explained. Thus began, for a human being for the first time, a remarkable surgical adventure, which is now replicated thousands of times throughout the world on a daily basis.

The chief resident in General Surgery at that time at the MGH, Dr. Ronald Malt, responded appropriately, meaning he assessed the possibility of replantation from the critical initial question "Should it be done?" before turning to the question of "How should it be done". Meanwhile the severed arm was wrapped in a sterile covering and immersed in a mixture of ice and sodium chloride solution.

In the absence of data from any prior replantations, the question of systemic risks to the patient from any adverse effects that might arise from toxic reactions generated by deteriorating tissues in the amputated extremity warranted serious reflection. This risk was particularly true of the renal risk from released myoglobin from degenerating muscle. In this regard the duration of the ischemia and the muscle mass involved were paramount. Based on the absence of other injuries, the relatively short time of ischemia before the likely resumption of circulation, the relatively small muscle mass of this amputated limb and prior experiences with successful "near amputation" cases, Dr. Malt elected to proceed. Particularly helpful was the observation that, despite this being an amputation caused by a railroad accident, the amputation was relatively free of major crushing of the remaining soft tissues. Tetanus toxoid was given, antibiotics were administered and the team effort began.

This first case accurately illustrated the marked importance of the optimizing effect of a counterintuitive series of the surgical sequences, specifically that a) the vascular repair should be deferred until a stable skeletal base had been established and b) that the restoration of channels for the venous return should be established before restoring arterial blood flow.

The wound was irrigated and extensively debrided, followed by fixation of the humeral shaft fracture using an intramedullary rod. Once this was secure, the vascular repairs restored venous circulation and then the arterial circulation. Nerve ends were identified and approximated but not repaired. Muscles were repaired without full closure of the skin, and the limb was encased in a plaster spica cast. A skin graft was applied to the partially closed wound on the fifth day. Importantly, infection had been avoided. This event generated a worldwide flood of press exposure.

Then followed multiple sequential surgeries and a prolonged rehabilitation, beginning with repair of the ulnar and median nerves at three months. The 4 cm gap in the musculocutaneous nerve was bridged using a brachiocutaneous nerve graft, and the 8 cm gap in the radial nerve was bridged by multiple grafts from the left femoral cutaneous nerve, despite a five-fold difference in cross sectional area between the proximal and distal ends. No radial nerve function recovered.

Parathesia in the hand appeared at a year, opponens function returned at 15 months and flexor sublimi and profundi function by 19 months, but no wrist function was recovered.

At 27 months an unusual tendon transfer was done to restore elbow flexion, powered by the pectoralis major. The tendon of the functionless extensor carpi radialis longus was freed distally, reversed in direction and the belly of the muscle was sewn to the biceps tendon at the elbow. The distal end of the tendon, now reversed in direction, was attached to the insertion of the pectoralis muscle, which had been detached from the humerus and mobilized. This transplant subsequently produced elbow flexion power which rated "good."

In the absence of any extensor motor function and with the partially recovered sublimis and profundus function occurring only as a mass response but not individually, plus the total lack of any other active motor function available at the wrist or hand, the unusual decision was made to fuse the wrist. Under these extreme circumstances, this combination of useful flexion power plus a wrist fused in the position of function provided him with the one essential motor function, a good gross grasp, but at the obvious price that he had to use his other hand to open his fingers.

After his intramedullary rod was removed because of painful crepitus in the shoulder, he entered into a fist fight, during which he refractured his humerus at the amputation site. This refracture healed without surgical intervention.

Similarly, at eight years after his amputation he entered into a fist fight again and fractured his wrist fusion, which also healed under conservative management.

When evaluated at eleven years post amputation, he was employed as an auto mechanic. The extensive calluses on his right hand spoke volumes about the utility of his replantation to his employment.

Forty-five years after the accident, he was living alone, fully independently, was on Social Security and had no long-term adverse effects of the replantations. Although at that time he was using the hand less because of his relatively undemanding existence, he had benefitted greatly from his replantation and the subsequent long series of operations compared to an amputation of the arm at the level of the deltoid insertion.

While the general publicity of this case was intense and worldwide, the replantation team made the specific decision not to publish a scientific report of this case until after a full year had passed. This was done in order to allow for the recognition of any adverse features, if any were to become manifest. However, because of the universal public press which the operation stimulated, many surgeons near and far adopted and repeated this original example of replantation of a fully amputated extremity.

Following this first case, it seemed prudent for us to plan thoroughly for the skeletal stabilization of any subsequent potential replantation opportunities at every level of the appendicular skeleton. I did this on cadavers at Harvard Medical School. Many of the solutions were obvious, as had been the use of the intramedullary rod in the first case. Intramedullary rods are excellent for this purpose, particularly because they restore stability without adding bulk, provide excellent and immediate immobilization of the fracture and are often appropriate for simultaneously shortening the limb. Shortening is often an asset because frequently the ends of the fracture can be shaped in ways to both increase the contact area and to control rotation, as well as the advantage that shortening provides aid in reapproximating vessels, nerves and muscles. Whenever possible, intramedullary fixation is preferred.

The more complex regions for stabilization were the most

proximal region of the humerus, the area of the elbow, and the region of the wrist. Studies were also done throughout the lower extremity, despite the clear decrease in motivation for lower extremity replantation that exists because of many factors, including the more limited function provided by the foot versus the hand, the longer path and lower success of the nerve repairs in the lower extremities and the availability of more successful prostheses for many lower extremity amputations.

The most innovative of these exploratory stabilizations that I created at that time was to address an amputation through the wrist. And it was not too long before an opportunity arose to put this solution to use. Our second case also illustrated many cardinal features of replantations.

This twenty-seven year old carpenter amputated his nondominant left hand through the wrist in a radial saw accident. With incredible equanimity he shut off his saw so he could safely retrieve his hand from the scrap box. When he showed his amputated hand and the stump of the wrist to his boss, the boss fainted. The carpenter then revived the boss, got him to apply a tourniquet to the amputated arm and to drive to the nearest hospital. This hospital turned out to be uniquely ill-equipped for this type of injury because it was a maternity hospital. There, the carpenter calmly requested sterile drapes to cover his injured wrist and hand and suggested that someone arrange for a police escort to the MGH, because he had heard about our first case. His injury met all the key criteria for the effort at a limb replantation.

This was done, using for the first time, the technique I developed on a cadaver in anticipation of an amputation through the wrist. To shorten the limb for easier approximation of the vessels and nerves, the proximal carpal row was excised, as well as some of the distal radius and ulna. Fixation at the wrist was achieved by using two crossed, stout, threaded Kirchner wires. The first was driven in through the length of the second metacarpal into the radius with the wrist in neutral flexion-extension. The wrist was then flexed to the position of function with this wire in place, bending the wire. The second wire was then driven across the wrist with the wrist in the dorsiflexed position, through the 5$^{th}$ intercarpal into the radius. This second wire, acting in conjunction with the now-deformed first wire, both immobilised the wrist area and locked the wrist in this optimized position of function.

At three months the severed flexor profundi tendons were freed from scar and reattached using sections of sublimi as tendon grafts. The severed ends of the ulnar and median nerves were approximated and sutured. Later on, the distal phalangeal joints of the fingers were fused in slight flexion.

When assessed at six years, he had quite useful flexion function, with protective sensation, two point discrimination of about 1 inch and excellent pinch. He used this hand extensively, including holding nails, lifting a glass and carrying heavy loads. He had returned to full labor as a carpenter.

His case illustrates the huge advantages of a replantation of a sensate and functioning hand over any prosthesis. It also reflected the inherent advantages that accrue to the patient from an amputation this far distally, including less muscle damage, preservation of many motors, shorter distance and time for nerve regeneration and the fact that the amputation occurs primarily through bone and tendons. Obviously under these conditions, many more options were available for motor restoration than in the first case and far fewer motor needs needed to be filled. It also illustrated the marked advantages of a sharp amputation with minimal damage to surrounding tissues.

These early experiments stimulated widespread new programs in virtually every aspect of this valuable and rewarding surgery, including microvascular techniques for restoration of circulation. Every day and every night around the world limb replantation now enriches the lives of thousands of patients.

Many of the early principles generated in these and other early experiences have weathered the challenges of time, including also the important but difficult requirement that the surgeons, the patients and the families also be prepared for failure of the replantations and the necessity for re-amputations if the limitations of the restoration prove to be insurmountable.

## SELECTED RELATED REFERENCES

Malt RA. "Clinical Aspects of Losing Limbs" in Welch, C.E. (ed.): Advances in Surgery, vol. 2, Chicago: Year Book Medical Publishers, Inc.,1966.

Malt RA, Harris WH. "Replantation of Limbs," in Najarian, J.S. and Simmons, R.L. (eds.): Transplantation, Philadelphia: Lea and Febiger, 1972.

Malt RA, McKhann CF. Replantation of Severed Arms. JAMA 189:114, 1964.

Ch'en CW, Ch'en Y-C, Pao Y-S. Salvage of the Forearm Following Complete Traumatic Amputation. Chinese Med. J. 82:632, 1963.

Malt RA and Harris WH. Replantation of Limbs. Somerville, NJ, Ethicon, Inc., 1965.

Malt RA, Remensnyder JP, Harris WH. Long-term Utility of Replanted Arms. Annals of Surgery 176: 334-342, 1972.

Harris WH, Malt RA. Late Results of Human Limb Replantation. J. Trauma, 14:44-52, 1974.

Malt RA, Harris WH. Replanted Amputated Arms. In: Management of Trauma, Ed. EF Cave, JF Burke, and RJ Boyd, Chicago: Year Book Medical Publishers, Inc. 1183-1193, 1974.

Malt RA. Clinical Aspects of Restoring Limbs, in Welch CE (ed.) Advances in Surgery, vol. 2, Chicago: Year Book Medical Publishers, Inc., 1966.

# CHAPTER 5

-- -- -- -- -- -- -- -- -- -- -- -- -- -- -- -- --

# CONQUERING FATAL
# PULMONARY EMBOLI

A DEVASTATING IMPACT ON MY life as a young orthopaedic surgeon was to influence my research and clinical career ever after, the death in my arms of a patient on whom I had just completed a hip reconstruction. The cause – a fatal pulmonary embolism. John was a "salt of the earth" guy. Married, two kids. Minimal education but dedicated to creating a family and an environment in which he could succeed and his kids could exceed his level of success. Snuffed out, suddenly, abruptly, just because of painful arthritis of the hip joint and snuffed out by a "bolt from the blue".

Certainly, from my medical education and orthopaedic training, I knew, conceptually, of fatal pulmonary emboli. This was different. This was the full emotional assault of the devastating effect of this "strange" pathology on an individual, on a family, on a patient of mine, and on me.

Before the requirements of time and my occupation as Chief Resident in Orthopaedic Surgery at the Massachusetts General Hospital forced me to move on and deal with the ever-intruding present, I made a quiet resolve to myself that I never wanted this to occur again to a patient of mine. Moreover, if possible, I, personally, would do something to reduce or stop fatal pulmonary emboli generally.

As reality set in, that highly-charged, emotional expression of grief looked more and more unlikely, even impossible – a pipe dream. No one, but no one, felt that fatal pulmonary emboli were the responsibility of orthopaedic surgeons. This poorly understood condition, if and when it was studied, was the purview of internists, hematologists, pulmonologists, general surgeons or pathologists. No work on the problem had ever been done by an orthopaedist. And what effort there was against fatal pulmonary emboli was sufficiently limited that the expression grew to be widely accepted that a fatal pulmonary embolus was, indeed, an "Act of God" or a "bolt from the blue."

Recall that in those days (the 1950s) the Homans sign was the common test during the physical examination for assessing the presence of a calf thrombus. It was felt to be positive in about 12% of postoperative adult hip patients. In stark contrast with that low figure, once venography was routinely used in research studies to assess the incidence of deep venous thrombosis (DVT) after total hip surgery, the true incidence in this group was shown to be 50%!

Just as the clinical diagnosis of DVT was severely underestimated, so too was the incidence of pulmonary emboli. Years later, our research established the incidence of pulmonary embolism following total hip replacement to be 23% (1), not rare and not a mysterious event. And the actual risk was certainly higher because in our study half of these patients were on pharmacologic prophylaxis against DVT.

Prophylaxis against DVT was virtually nonexistent, despite being an era when prolonged bed rest after major surgery, especially after lower extremity surgery, was the mode. If the diagnosis of pulmonary embolism was made, the anticoagulants were added after the fact.

The "Trendelenburg" operation – an extremely high risk attempt to remove the embolus from the lung in desperately ill patients who had had a major or massive pulmonary embolus – was rarely used and generally fatal.

In short, venous thromboembolism was poorly understood,

underdiagnosed, and without effective prophylaxis. Treatment was after the fact by anticoagulation.

This high level of ignorance severely compromised patient survival when total hip surgery overtook the field of reconstructive surgery of the hip. The arresting figures for fatal pulmonary emboli among total hip patients was 2% (See Table 1). This meant that one of every 50 patients receiving a total hip for the treatment of hip pain caused by a benign condition, arthritis of the hip, would die from a fatal pulmonary embolus.

Table 5.1 Incidence of Fatal Pulmonary Emboli (FPE) During Early Years of THA

| Author | $N$ | FPE |
|--------|-----|-----|
|        |     |     |
| Coventry et al. (2) | 2012 | 3.4% |
| Bombelli et al. (3) | 300 | 2.3% |
| Johnson, Green, Charnley et al. (4) | 1174 | 2.3% |

That is a terrible price to pay in the management of a benign condition. When we started doing total hip surgery in the United States in 1969, my prior apprehension about fatal pulmonary emboli was sharply magnified, particularly because of the truly spectacular clinical results of total hip surgery among those who escaped serious complications. What to do?

Fortunately Ed Salzman, a clinician-scientist general surgeon interested in bleeding and clotting issues, Roman de Sanctis, a cardiologist, and I had already shown both in hip fracture patients over the age of 74 years (5) and in hip reconstruction patients over 40 years of age undergoing cup arthroplasty (6) that prophylactic anticoagulation using Warfarin was both safe and effective in prospective controlled randomized studies reducing DVT and pulmonary emboli.

These landmark articles were the first two to be published in the Western Hemisphere that established that a pharmacological agent could effectively reduce the risks of DVT and pulmonary emboli among adult patients undergoing reconstructive hip surgery or hip fractures. Importantly they initiated the use of prophylaxis against venous thromboembolic disease (VTED) in orthopaedic populations in the USA and Canada. These two publications overcame the uncertainty of the previously unknown risk-reward ratio of adding anticoagulants simultaneously with major hip surgery in these two groups. While of value by proving effective and safe prophylaxis, they were even more powerful in stimulating parallel efforts to improve on their success in reducing postoperative venous thromboembolic disease after adult hip surgery.

The magnitude of the problem was succinctly presented in the Hip Society Award article entitled: "New advances in the prevention, diagnosis, and cost effectiveness of Venous Thromboembolic Disease in Patients with Total Hip Replacements " (7): as follows "In the early days of total hip replacement (THR) surgery, fatal pulmonary embolism (FPE) was the leading cause of mortality." As shown in Table 1 Series from Charnley's group, Bombelli et al. (4) and the Mayo Clinic (2) reported an incidence of fatal PE of 2.3%, 2.3% and 3.4%, respectively, in unprotected patients. Even as late as 1985 Kakkar and associates (8) have reported a similar incidence of fatal PE in unprotected patients undergoing THR surgery: 2.3%.

Based on our first two studies in 1966 and 1967 (5, 6) we proposed Warfarin prophylaxis for hip surgery in the older adult patients. They were followed by studies evaluating the various anti-platelet agents including our innovative introduction of the antiplatelet agent aspirin (9) and subsequently the application of those types of pharmacologic prophylaxis to the management of patients undergoing total hip replacements (10). All these investigations of the efficiency of aspirin as a prophylactic method following hip surgery in the adult, as well as additional studies by other investigators, effected a revolution in clinical practices in the USA among surgeons doing total hip replacements.

Subsequent studies (11) showed not only the dramatic improvement in safety by adopting the low dose Warfarin regimen advocated by

Hull et al. (12) but also demonstrated, remarkably, that the incidence of fatal pulmonary emboli could be reduced from 20 per 1000 patients to three per 1000 patients and, equally remarkably, this stunning improvement in saving lives was, indeed, cost effective compared to no prophylaxis, independent of the huge benefit from the lives saved.

Strikingly, the acceptance of this radical change in management of total hip surgery patients rose rapidly. By 1985, at a time when 25% of responders to a nationwide survey among total hip surgeons had acknowledged having at least one patient in the last five years die from a fatal pulmonary embolism, the incidence of hip surgeons using some form of prophylaxis had risen to 84%. That figure continued upward, and by 1990, it was 90% (11). This is a remarkably dramatic and major change nationwide in surgical practice in the short 20 years since total hip surgery was permitted in the United States by the FDA.

Continuing progress has substantially advanced the field. A major thrust was to stratify the risk so as to specify subsets of the patient population which could effectively be protected with the least-risk modality. Another major effort was toward reducing the bleeding complications. Because of simplicity of use and low risk of bleeding complications, many surgeons have elected to rely on aspirin, an agent which we were the first to introduce (9).

Of great import also was the striking progress made in improving the efficacy of external pneumatic compression devices (13). In addition to their strong appeal of not relying on pharmacological alteration of the bleeding-clotting homeostasis and thus the zero increase in bleeding risk, sophisticated changes in the device to correlate the pressure impulse with the respiratory cycle dramatically increased the efficiency. Now, mobile compression devices have been shown to be both extremely efficacious (13,14,15) and safe as well as cost-efficient and to do this for both THA and TKA patients. See Table 5.2.

Table 5. 2 Incidence of Fatal Pulmonary Emboli During Current THA Practice

| Author | $N$ | FPE | FPE% |
|---|---|---|---|
| Colwell et al. (15) | 1509 | 0 or 1* | 0 or 0.06 |
| Colwell et al. (16) | 414 | 0 | 0 |
| Ericksson (17) | 3153 | 6** | 0.2 |

*Only one patient died in this series. No autopsy was done. So either 1 or none had a fatal pulmonary embolus.

**6 total deaths from 3153 patients, all on either rivaroxaban or enoxaparin. The cause of death was not reported but even if the unlikely situation was that all 6 were caused by fatal pulmonary emboli, this would be only 0.2%.

Several studies show the value of preoperative risk stratification in terms of the efficacy of the prophylaxis and reduction in the complications, including prosthetic joint infection, even though the population in these studies and the criteria for risk stratifications differ (18,19). In addition to the valuable role that aspirin prophylaxis in standard doses played in protecting those low risk patients, others have presented data supporting the efficacy of low-dose aspirin (81 mg) over 325 mgm doses in low-risk groups (20) and that aspirin may even be safer and more effective for even the higher risk group (21). In another study using standard dose aspirin plus mobile compression boots in the low-risk group of primary and revision and surface replacement THA patients had an incidence of any clinically apparent VTED within six weeks of just 5 cases/1000 plus having lower wound complications, major bleeding complications and readmission within six months than the higher risk group given Warfarin (22). In the report by Huang et al. (21) aspirin was as effective as Warfarin even in the high risk patients receiving either primary or revision TKA or THA with fewer complications.

Another presentation suggested that aspirin treatment of subpopliteal detected venous thrombli was safe as well (23).

Interestingly, also, the analyses by Della Valle et al. (24) reported that of 269 patients with documented pulmonary emboli shown on

CTA, only two died and neither the location nor unilateral versus bilateral emboli influenced the fatal outcome.

## SELECTED RELATED REFERENCES

1. Harris WH, McKusick K, Athanasoulis CA, Waltman AC, Strauss HW. The Detection of Pulmonary Emboli after Total Hip Replacement Using Serial C1502 Pulmonary Scans. J. Bone Joint Surg., 66-A: 1388-1392, 1984.
2. Coventry MB, Nolan DR, Bechenbaugh RD. Delayed prophylactic anticoagulation in 2,012 total hip arthroplasties. J. Bone Joint Surg. 55(A): 1487, 1973.
3. Bombelli R, Gerundini M, Aaronson B. Early results of the RM isoelastic cementless total hip prosthesis: 300 consecutive cases with 2-year follow-up. The Hip Proceedings of the Twelfth Open Scientific Meeting of the Hip Society, St. Louis, 1984 The C.V. Mosby Co. 133-145, 1984.
4. Johnson R, Green JR, Charnley J. Pulmonary embolism and its prophylaxis following Charnley, total hip replacement. Clin Ortho Rel Res (127):123-132, 1977.
5. Salzman EW, Harris WH, DeSanctis RW. Anticoagulation for Prevention Thromboembolism Following Fractures of the Hip. N. Engl. J. Med., 275:122-130, 1966.
6. Harris WH, Salzman EW, DeSanctis RW. The Prevention of Thromboembolic Disease by Prophylactic Anticoagulation. J. Bone Joint Surg., 49-A: 81-89, 1967.
7. Paiement GD, Bel D, Wessinger SJ, Harris WH. The Otto Aufranc Award Paper – New Advances in the Prevention, Diagnosis and Cost-Effectiveness of Venous Thromboembolic Disease Patients with Total Hip Replacement. In: The Hip, Proceedings of the Fourteenth Open Scientific Meeting of the Hip Society, CV Mosby Co., Ed. RA Brand, St. Louis: pp.94-119, 1986.
8. Khakar VV, Foke PJ, Murray WJG, Pals T, Merenstein D. et al. Heparin and dihydroergotamine prophylaxis against

thromboembolism after hip arthroplasty. J Bone Joint Surg 67(B): 538, 1985.

9. Salzman EW, Harris WH, DeSanctis RW. Reduction in Venous Thromboembolism by Agents Affecting Platelet Function. N. Engl. J. Med., 284:1287-1292, 1971.

10. Harris WH, Salzman EW, DeSanctis RW, Coutts RD. Prevention of Venous Thromboembolism Following Total Hip Replacement. JAMA 220: 1319-1322, 1972.

11. Paiement GD, Beisaw NE, Harris WH, Wessinger SJ, Wyman E. Advances in Prevention of Venous Thromboembolic Disease after Elective Hip Surgery. In: Instructional Course Lectures, Vol. XXXIX, 413-421, 1990, Walter B Greene, editor. Chicago: American Academy of Orthopedic Surgeons, 1990.

12. Hull R, Hirsh J, Jay R, Carter D, England C, Gent M, Turpie AGG, Mc Laughlin D, Dodd P, Thomas M, Raskala G, Okeford P. Differential intensities of oral anticoagulant therapy in the treatment of proximal vein thrombosis. N. Eng. J. Med. 207:1676, 1982.

13. Colwell CW Jr, Froimson MI, Ritter MA, Trousdale RT, Buehler KC, Spitzer AI, Donaldson TK, Padgett DE. Thrombosis prevention after total hip arthroplasty: a prospective, randomized trial comparing a mobile compression device with low-molecular-weight heparin. J Bone Joint Surg (Am) 92(3): 527-535. doi: 10.2106/JBJS.I.00047, 2010.

14. Falck-Ytter Y, Francis CW, Johanson NA, Curley C, Dahl OE, Schulman S, Ortel TL, Pauker SG, Colwell CW Jr. American College of Chest Physicians. Chest 141(2 Suppl): 278S-325S. doi:10.1378/chest.11-2404, 2012.

15. Colwell C.W Jr, Froimson, MI, Anseth SD, Giori NJ, Hamilton WG, Barrack RL, Buehler KC, Mont MA, Padgett DE, Pulido PA, Barnes CL. A mobile compression device for thrombosis prevention in hip and knee arthroplasty. J Bone Joint Surg (Am) 96(3):177-183. doi: 10.2106/JBJS.L.01031, 2014.

16. Colwell C Jr, Froimson MI, Mont MA, Ritter MA, Trousdale RT, Buehler KC, Spitzer AI, Donaldson TK, Padgett DE. Cost-effectiveness of venous thromboembolism prophylaxis with a new mobile device after total hip arthroplasty. J Arthroplasty 27(8): 1513-1517.e11. doi 10.1016/j.arth.2012.03.024 Epub 2012.

17. Eriksson BI, Friedman RJ, Haas S et al. Rivaroxaban versus enoxaparin for thromboprophylaxis after hip. N Engl J Med. 358(26):2765-2775, June 2008.

18. Tan TL, Maltenforr M, Chen AF, Shahi A, Higuera CA, Siqueira M, Hansen EN, Sing D, Parvizi J. Efficacy of Venous Thromboembolism Prophylaxis in Total Joint Arthroplasty Based on Risk Stratification: Should Potent Anticoagulation Be Used in Higher Risk Patients? Poster Number 104 American Association of Hip and Knee Surgeons Dallas Texas November 10-13, 2015.

19. Doran JP, Odeh KI, Szulc A, Murphy H, Smith D, Bosco JA, Iorio R. Risk-stratified VTE Prophylaxis Following TJA: Aspirin and SPCDs vs. Aggressive Chemoprophylaxis. Poster Number 106 American Association of Hip and Knee Surgeons Dallas Texas November 10-13, 2015.

20. Parvizi J, Huang R, Restrepo C, Chen A, Austin M, Hozack W, Lonner J. Low Dose Aspirin is an Efficient Chemoprophylaxis for Venous Thromboembolism Following Total Joint Arthroplasty: An Interim Analysis. Poster Number 103 American Association of Hip and Knee Surgeons Dallas Texas November 10-13, 2015.

21. Huang R, Parvizi J, Hozack W, Austin M. Aspirin is as Effective as and Safer than Warfarin for Patients at Elevated Risk of Venous Thromboembolism Undergoing Total Joint Arthroplasty. Poster Number 102 American Association of Hip and Knee Surgeons Dallas Texas November 10-13, 2015.

22. Nam D, Nunley RM, Johnson SR, Keeney JA, Clohisy JC, Barrack RL. Use of a Risk Stratification Protocol for Thromboembolism Prophylaxis Following Hip Arthroplasty.

Poster Number 107 American Association of Hip and Knee Surgeons Dallas Texas November 10-13, 2015.

23. Parcells BW, Hartzband MA, Levine HB, Klein GR. Propagation of Infrapopliteal Venous Thrombotic Events with Aspirin Treatment Following Total Knee Arthroplasty. Poster Number 109 American Association of Hip and Knee Surgeons Dallas Texas November 10-13, 2015.

24. Della Valle AG, Perez AB, Lee Y, Konin G, Endo Y, Saboeiro G, Sharrock N, Salvati E. The Clinical Severity of PE Following Joint Replacement is Unrelated to the Location of Emboli in the Pulmonary Vasculature. Poster Number 108 American Association of Hip and Knee Surgeons Dallas Texas November 10-13, 2015.

# CHAPTER 6

# EXPANDING ANGIOGRAPHY

THE DRAMATIC EXPANSION OF ANGIOGRAPHY aided my research in many ways. For example, as presented in Chapter 5, it was crucial in establishing the full incidence of venous thromboembolism disease following total hip arthroplasty (THA). Just as venography established that the incidence of venous thrombosis in the elderly THA population was 50%, it was pulmonary angiography that provided the rigorous data establishing that the true incidence of pulmonary emboli following THA was at least 23%. These angiographic data, for the first time, created the full picture of the prevalence of venous thromboembolic disease after THA.

From my close association with the Vascular Radiology Department based on these angiographic studies, grew an expanded application of angiography on a surprising range of cases unrelated to venous thrombosis disease at all. Three cases will demonstrate that useful and unusual interaction.

RC was an overweight 49 year old man with low physical activity levels and high beer consumption levels. In addition to his severe osteoarthritis of the hip, his preoperative medical assessment revealed both marked obesity and unyielding hypertension. During his total hip operation under general anaesthesia his ECG showed the sudden onset of a severe disturbance. Our cardiac consultant interpreted this event as a massive myocardial infarction, with a high risk of being fatal.

Since the total hip procedure had been nearly completed when this occurred, immediate steps were taken to finish the operation expeditiously and rapidly close the wound. But what should be the next step? Our cardiac consultant, evaluating all the factors involved, recommended that if he were not in the operating room, that is, if this infarction had occurred spontaneously without the added factor of occurring during hip surgery, he would recommend an immediate coronary angioplasty.

I responded that, if that is the optimum course, let's do it. We should proceed immediately to the insertion of a coronary artery stent to preserve myocardial function, prevent further damage and not wait for further evaluation of this infarction over the several hours necessary to allow him to recover from general anesthesia because of all the risks which that course entailed. Thus began the unusual sequence of an immediate coronary angioplasty while still under continuing anaesthesia from a total hip operation.

Central to this decision on my part was the deep trust I had in our cardiologist, generated over many years of his providing insightful, daring and correct advice in a series of difficult, dramatic experiences during emergency cardiac and pulmonary episodes, both in and out of the operating room.

With the patient still on the operating table, after rapidly closing the hip wound, we commandeered an elevator and proceeded out of the operating room, into the elevator, and through the corridors of the hospital with a full crew of anesthesiologists, attendants, surgeons and a cardiologist in tow, all the way to the cardiac catheterization suite. There, still under anaesthesia, the catheterization and insertion of the stent proceeded successfully.

The one huge disadvantage of this unorthodox move was the requirement that after the coronary stent procedure, it was imperative immediately to anticoagulate the patient to protect the stent from clotting. Not only did this mean full anticoagulation, it meant doing so with heparin. While essential for the stent, heparin anticoagulation was a dire threat to his large, fresh hip wound, because of the massive increase in wound bleeding it would cause. Briefly stated, we had to balance the calculated risk of death from this massive myocardial infarction versus the severe bleeding that would occur in the hip wound. This bleeding could well disrupt the wound, which in turn would likely lead to deep infection in the hip. As disastrous as deep infection in the hip would be for this obese man with a fresh total hip and a fresh myocardial infarction, the extreme severity of his infarction took precedence. So we proceeded with both the stent and the heparin.

The results were remarkable on both scores. The stent was very effective in restoring cardiac circulation and because his myocardial damage was minimized, he survived this massive infarction.

And equally so, the hip wound did well despite the massive bleeding into the wound. The bleeding was so extensive that he required nine blood transfusions. The hematoma produced a vast black and blue discoloration of his flank from the severe bleeding, extending from his armpit to his ankle. This reflected the amount of blood released into his wound because of the anticoagulation.

But neither did the wound break down nor did his wound become infected. A major factor in the stability of the wound was the unusual technique of wound closure I used to approximate the wound in the shortest possible time. Instead of closing the wound meticulously layer by layer as is always recommended, expeditiously the wound was closed using multiple, massive sutures delivered in the form of large encompassing sutures, placed across the entire wound using those large needles customarily applied by pathologists to close the wounds of an autopsy. These were applied in the form of "retention sutures". They could, and did, withstand all the disruptive hydrostatic forces from his extensive wound bleeding.

In fact, his cardiac situation stabilized well enough and his hip wound was secure enough that on the 9th postoperative day he was discharged to his home.

The second example of unusual and innovative use of angiography was the result of an attempt to reconstruct the hip of a patient whose total hip reconstruction aimed to correct a complete developmental dislocation of the hip, addressing the problem and using techniques described in Chapter 10. In this second example, the angiography was needed to identify the obscure source of the unrelenting intraabdominal bleeding artery and then to stop the bleeding.

This 44-year-old woman who had had prior intraabdominal surgery presented with severe osteoarthritis of the right hip secondary to severe developmental dysplasia. Because of the major deficiency of the periacetabular bone which resulted from the marked underdevelopment of her developmental dysplasia, I needed to increase the volume of bone available to reconstruct her hip socket. To do this I had to bolt her amputated femoral head to the wing of the ilium as a graft. In order to do this, it was necessary to expose the <u>inside</u> of the pelvis, to be able to bolt the femoral head graft to the very thin pelvis.

To create this extensive exposure inside the pelvis, the soft tissues along the inside of the pelvic bone at the region of the acetabulum are stripped off the inside of the pelvis. This was done blindly by using my index finger which was inserted inside the pelvis through the lateral hip wound. While I had developed this technique and had studied it thoroughly before ever using it in a patient and, moreover, had used it in many patients previously without difficulty, in this patient major bleeding resulted as that area was being exposed. The bleeding was probably related to scar tissue that developed following her prior intraabdominal surgery.

All my extensive efforts to control the bleeding were unsuccessful. I then sought assistance from a skilled general surgeon who was very familiar with surgery in this region of the abdomen. Despite all his efforts the bleeding continued unabated.

The customary next step would be to pack the area of bleeding with gauze packs to temporarily control the bleeding, finish the hip operation, roll the patient off her side onto her back and open the abdomen for direct visualization through the abdominal incision to control the bleeding. All of these efforts carried serious risks, including a substantial increased risk of deep infection.

The unique alternative I adopted was to finish the hip operation rapidly and take the patient while still on the operating table and still under anesthesia through the halls of the hospital to the angiography suite in radiology, for angiography.

Although highly unorthodox and the first time such a venture had been done, it worked just as desired. Under radiographic control, the bleeding artery was identified and then a physician-created "artificial blood clot" was made from non-absorbable sterile plastic material which was delivered by the catheter directly to the bleeding site. It occluded the rent in the artery. The bleeding stopped. The hip reconstruction succeeded. The wound healed well and when last examined 12 years later, her outcome was highly successful.

The third example represents yet another innovative use of an innovative technique. These angiographic techniques had been devised to solve many different problems. For example, angiography could permit visualization of the coronary arteries, demonstrate pulmonary emboli, and stop bleeding in awkward locations or under situations of bleeding that were extremely difficult to control, as just demonstrated above. It is used widely in such cases as severe trauma, etc. But in this case we used it to prevent the bleeding rather than to stop it.

CA was a 43-year-old man who presented with a metastatic, destructive, malignant lesion in his femur. But since all of his other tests were negative, the primary source of his cancer was unknown. This situation required a biopsy of the lesion to establish the diagnosis. While for most of these cases a needle biopsy is a standard and safe procedure, the appearance of this lesion in this patient suggested that the lesion was likely to be clear cell carcinoma of the kidney. Often these tumors are so vascular that the bleeding, even from a needle biopsy, can be excessive and continuous to the point of requiring an open operation simply to control the bleeding from the needle puncture. And that operation is, in itself, quite challenging due to the risk of more bleeding.

For this patient we elected to use angiographic techniques to block the major blood supply to this potentially extremely vascular lesion on a prophylactic basis. An angiogram, done to identify the location and extent of the blood supply to the lesion, showed the vasculature to be richly overdeveloped. This, in itself, was a further suggestion of the etiology.

Prior to the needle biopsy, the arteries directly leading to the lesion were occluded. The biopsy confirmed that the lesion was a clear-cell carcinoma of the kidney, and occurred without any unusual bleeding or subsequent complications.

## SELECTED RELATED REFERENCES

Harris WH, Crothers OD, Oh I. Total Hip Replacement and Femoral-Head Bone-Grafting for Severe Acetabular Deficiency in Adults. J. Bone Joint Surg., 59-A:752-759, 1977.

Harris WH, Crothers OD. Reattachment of the Greater Trochanter in Total Hip Replacement Arthroplasty. J. Bone Joint Surg., 60-A:211-213, 1978.

Athanasoulis CA, Harris WH, Stock JR, Waltman AC. Arterial Embolization to Control Pelvic Hemorrhage. In: The Hip, Proceedings of the Seventh Open Scientific Meeting of the Hip Society. Ed. CB Sledge, St. Louis: CV Mosby Co., pp. 247-259, 1979.

Stock JR, Athanasoulis CA, Harris WH, Waltman AC, Novelline RA, Greenfield AJ. Transcatheter Embolization for the Control of Wound Hemorrhage Following Hip Surgery. J. Bone Joint Surg., 62-A: 1000-1003, 1980.

Stock JR, Harris WH, Athanasoulis CA. The Role of Diagnostic and Therapeutic Angiography in Trauma to the Pelvis. Clin Orthop Rel Res. 151:31-40, 1980.

Harris WH. Allografting in Total Hip Arthroplasty: In Adults with Severe Acetabular Deficiency Including a Surgical Technique for Bolting the Graft to the Illum. Clin Orthop Rel Res. 162:150-164, 1982.

Harris WH, McKusick K, Athanasoulis CA, Waltman AC, Strauss HW. The Detection of Pulmonary Emboli after Total Hip Replacement Using Serial C1502 Pulmonary Scans. J. Bone Joint Surg., 66-A: 1388-1392, 1984.

# CHAPTER 7

## STEM FRACTURES

THREE FACTORS CONTRIBUTED TO THE arresting problem of fracture of the femoral stems in the early days of total hip replacement: poor design of the stems, weak metallurgy of the stem, and failure of cement fixation. Generally stem fracture constituted an abrupt clinical failure with sharp pain and acute loss of function. But occasionally the patient would be able to continue with somewhat compromised function, even for years, before reoperation was required.

These fractures were fatigue failures of the metal stem. The basic requirement for the fatigue cycling was the failure of the cement fixation proximally while distal fixation remained intact. Because of the offset loading of the hip, varus strains led to fracture of the stem initiating primarily laterally but also commonly with a rotary component because of the load on the femoral head tending to spin the stem into internal rotation, particularly during stair climbing and rising from a chair.

It was also true that these three issues, stem design, weak metallurgy and failure of fixation, were intimately interrelated. Many early stem designs followed the profile of the Thompson or Moore hemiarthroplasty prostheses which had, after all, been planned for press fit application, not cement fixation. As a result many of the early designs had narrow medial borders, sharp corners and relatively thin lateral contours. They also were not necessarily designed with optimal configurations to offset internal rotation torque. Consequently these designs produced high stress concentrations in the cement, particularly at the regions of the sharp corners on the narrow regions. These inappropriate stem design features promoted fatigue failure of the cement, particularly proximally, but also especially in the presence of flaws in the cement mantle or defects in the cementing techniques. Moreover, in the early days the femoral stems were cast, not forged. Therefore, since many patients take 2 million steps a year, all these features contributed to fatigue fracture of the stem, which could occur in the early years after insertion.

Fortunately this complication has been, for all intents and purposes, eliminated. Each of the three contributing factors has been dramatically improved. The chapter on cemented femoral stem design outlines those improvements which we innovated in femoral stem design. The subsequent change throughout the industry to using forged materials was very effective. And the chapter on femoral cementing techniques displays the wide range of improvement we and others advanced in that realm. Thus, an alarming and abrupt failure mechanism of the early days of cemented femoral stems has succumbed to these advances in cemented femoral component design and materials, and cementing techniques.

# CHAPTER 8

- - - - - - - - - - - - - - - - - -

# FIXATION OF COMPONENTS
# TO THE SKELETON

## CEMENTED FIXATION

OVER ITS INITIAL FOUR DECADES, a prime limitation of total hip replacement surgery was the key problem of obtaining and maintaining fixation of the components to the skeleton. And yet despite this limitation in obtaining and maintaining fixation of the components to the skeleton using cement, cement fixation was absolutely critical to both the survival and then the flourishing of the concept of total hip replacement.

It is of special interest, but perhaps not completely surprising, that every inventor of total hip arthroplasty from Gluck (1) to Charnley started with press fit concepts. As an aside, however, Gluck also reflected his extremely broad innovative spirit by also experimenting with a rapid-curing bonding agent. In Charnley's desire to find a method to replace "press fit" fixation of both acetabular and femoral components for his strikingly creative operation, he approached Dennis Smith, a materials scientist in the Turner Dental Hospital in Manchester for advice. (2)

His question was along the lines of "what material could be used in the body to provide fixation of a component of a radical new artificial joint to the upper portion of the femur?" Smith misunderstood the question. To him, being "in the mouth" was the same as being "in the body", which it surely is not. External dental devices residing "in the mouth" are simply not subject to the challenging internal environment of being "in the body." His reply was "methyl methacrylate", referring to the pink-colored acrylic plastic that had widely successful use in the mouth as portions of bridges and other artificial dental devices.

Fortunately, Charnley's adoption of Smith's incorrect answer serendipitously launched total hip surgery. This event is a special and yet curious example of Pasteur's motto that "chance favors only the prepared mind." Had Charnley's mind not been prepared by his quest for a "total hip replacement", the question would never have been asked of Dennis Smith and, thus, even the incorrect reply would not have occurred. But it did. And not only did it enable total hip surgery to be born, this same approach, i.e., the use of "bone cement" for fixing components of total hips and total knees to the skeleton, is still in use, five decades later.

But, cement fixation had extensive limitations, including such major flaws as resulting in hypotension and even death, promoting periprosthetic osteolysis, leading to stem fractures, and allowing linear osteolysis to progressively loosen many cemented acetabular components, all in addition to the all too common failure of fixation of femoral components.

Hypotension was an alarming response to the use of bone cement in surgery. As a result of the effects of the monomer, marked hypotension including sudden death could occur, particularly in the elderly, and especially if they were hypovolemic to start with. This complication, once recognized as being caused by the monomer, could be nearly universally avoided by accurate volume replacement prior to the use of the bone cement. Thus, the major subsequent endeavors to improve cement fixation were aimed at preventing loosening and to avoid lysis.

Failure to obtain femoral component fixation using cementing techniques during the early years of THA is clearly shown by the reports of Gruen et al. (3) and of Bechenbaugh and Ilstrup (4). See table 8.1.

Table 8.1 Loss of Fixation of Cemented Femoral Stems in the Early Years of THA

|  | N | Years | % Loose |
|---|---|---|---|
| Gruen, McNeice, Gregory, Amstutz (3) | 389 | 3 | 19.5 |
| Bechenbaugh, Ilstrup (4) | 333 | 4-7 | 24 |

Similarly, loosening rates of cemented acetabular components were high after 10 years and continued to increase with time despite a variety of design and technique changes, except in those patients age 75 and older. Protection against socket loosening in this group occurred primarily because of their limited longevity and reduced activity. But by 35 years after a cemented THR, over half of the cemented acetabular implants had been removed in patients under 50 years of age at the time of surgery, primarily for loosening, wear and lysis (5) (6).

Loose femoral components were commonly both painful and functioned poorly. An interesting contrast existed for loose cemented acetabular components in that very many of them escaped both of these outcomes, and could, for many years, continue to serve well without pain. Nevertheless, well-fixed implants were clearly superior and were the desired end.

Nevertheless, special additional credit must be given to Charnley not only for addressing the idea of methyl methacrylate as a grout for fixation of his remarkable surgery but also for the high degree of success he developed with his technique. In reality, however, many other surgeons were less successful – which ultimately prompted the massive efforts in two very different directions, to a) improve cement fixation and b) to eliminate cement fixation.

The adventures in improving cemented fixation of total hip components can be considered under seven headings, to all of which the Harris Orthopaedic Lab contributed substantially:

1. Improving the cement
2. Advancing the delivery process

3. Preparation of the femoral canal
4. Reducing the stress on the cement
5. Adding antibiotics to the cement
6. Defining the unique role of periprosthetic osteolysis in acetabular loosening of cemented sockets
7. Developing a method of grading the quality of femoral cementing

Of the many attempts to strengthen methyl methacrylate by adding other compounds, none have succeeded. Conversely, our innovation of removing something substantially added to its durability – namely decreasing its porosity (7). The mixing process of combining the powder with the liquid catalyst (the monomer) to obtain the useful putty-like consistency for delivery into the skeleton inevitably generated pores in the cement, seriously reducing its fatigue resistance. The Harris Orthopaedic Lab identified the importance of this feature and followed on with a solution by centrifugation (7-15). Others later developed vacuum mixing for the same purpose.

In addition, on the issue of strength of bone cement (16), we established that the ill-fated efforts to improve the distribution of the cement to the skeleton by using preparations of methyl methacrylate which had lower viscosity were compromised by its lower fatigue life. This disadvantage was compounded by our further observation that the low viscosity cement also had an increased porosity (17, 18).

As a result of all these investigations of methyl methacrylate itself, the properties of the cement were optimized, not by additives, but by porosity reduction. Also, notably, during these years a proposed alternative form of acrylic grouting material was clearly shown by Malchau and co-workers to be vastly inferior (19).

Major developments from the Harris Orthopaedic Lab also resulted in advancing the delivery process, for both femoral cementing and acetabular cementing. On the femoral side, in contrast with Charnley's "thumb packing" technique, we improved delivery by a series of steps consisting of a) inventing the medullary plug syringe to permit plugging the femoral canal (20-24), with a bolus of methyl methacrylate, b) centrifugation (7-15), c) delivering the pore-free cement into the femoral canal retrogradely from distal to proximal using a cement gun (22) and d) pressurizing the cement mass for both the femoral cementing (23-27) and the acetabulum (28) (see below).

Plugging the canal prior to delivering the cement had two great advantages. First, it limited the delivery of the cement to the proximal portion of the canal. Prior to this technique, often cement was pressed far beyond the area needed simply to enclose the femoral strain, making any subsequent femoral revision operation far more complex. Secondly, the plug allowed pressurization of the cement to be far more effective.

Our cement plug syringe enabled both accurate location of the plug and the utilization of only a half package of cement to create the plug. The plug syringe was calibrated for length so the plug could be delivered at the exact location required for the length of the stem used. While a variety of designs of preformed plastic plugs followed, all created to provide the same advantage, the cement plug was more reliable and could specifically be used to occlude the femur below the isthmus for plugging the canal for use with longer cemented stems, a feature not possible using preformed polymer plugs.

Delivery of the cement by our cement gun was also a major step forward. Firstly, it eliminated all those seams in the cement caused by repeated thumb packing, and especially the weakness of the cement mass that resulted from the inadvertent inclusion of blood into the layered cement between the "thumb thrusts". Moreover it delivered the cement retrogradely from the plug proximally, not forcing it from proximal to distal into an air pocket. Additionally the delivery of cement from distal to proximal in the femur often drove residual remaining blood and/or bone marrow proximally, up and out of the femur ahead of the advancing cement mass where it could be suctioned off.

Pressurization of the femoral cement (25-28) enhanced the cement fixation crucially, because it forced the cement deeply toward the cortex and into any remaining cancellous bone, maximizing the intrusion of cement into the interstices of the trabecular bone adjacent to the cortex. This intrusion was central to optimize the interdigitation that strengthens the lock of the cement-bone interface.

Following the demonstrated advantage to femoral cementing of pressurization of the femoral cement to produce greater and more uniform interdigitation with the host bone, we also generated similar devices to pressurize the acetabular cement (28). These consisted of seals of increasing size which occluded the opening of the acetabular recess, including the defect in the acetabular rim leading into the Haversian recess. These seals fit over the end of the nozzle of the cement gun during the injecting of the acetabular cement, effectively occluding the mouth of the acetabular recess. In this way positive pressure was applied and maintained, driving the cement deeply into the recesses in the bone.

Important further improvements by others focused on the preparation of the femur itself to receive the cement. They included the development of the water pic to wash out debris, blood and marrow and other techniques to reduce the contamination of the cement-bone interface by blood through the local use of adrenalin-soaked sponges. We and others advanced the idea of the systemic reduction of bleeding into the bed prepared for cement by using induced hypotensive anesthesia (29-31).

One additional aspect of the fixation of cemented femoral stems involved contributions from the Harris Orthopaedic Laboratory also, namely precoating the femoral stem with a thin layer of methyl methacrylate at the factory (32-48). This concept was advanced by Dr. Jo Miller (39-40). We put it to application with the "Precoat stem". Without doubt, any cement-stem interface prepared under manufacturing conditions has uniformity, freedom from imperfections and strength that cannot be achieved in the operating room. This was demonstrated at the clinical level by our use of the Precoat stem (35, 36). But the clinical experience elsewhere was mixed (37, 38), in part because of confusion about the size of some rasps to use with the Precoat stem, leading some surgeons to generate cement mantels that were too thin and thus subject to premature failure because of increased stresses.

We developed the first technique for measuring directly the strain in femoral bone cement under simulated human loading conditions (40-43) and also studied extensively both the initiation of cement failure and the failure mechanisms (44-54). These mechanisms, which included debonding (48-49), pores at the interface (54), and fatigue failure led to fragmentation (55). Critical to increased strain was a thin cement mantle (56, 57). To reduce those high strains near the tip of the cemented stem and overcome the frequent lack of centralization of the tip of the stem, we developed new designs for the tip of the stem and corresponding techniques for centralization of the tip (58, 59).

Our contribution to protecting the cement from excessive stress and thus from fatigue failure had two arms, a) the optimization of the design of the femoral component in order to reduce peak stress and b) intensive studies measuring the cement strains under simulated gait loading conditions using stress gauges built into the cement mantle. Those features of our extensive work on the design of the femoral component are contained in the section called "Cemented Stem Design" in Chapter 9.

In summary, as a result of all these activities the Harris Orthopaedic Laboratory made substantial advances for improving cement fixation on the femoral side at the cement-bone interface, within the cement itself, and at the cement-stem interface.

The improvements in design and materials in femoral stems augmented the advances in cementing techniques, which led to major reductions in loosening and revision for aseptic loosening of cemented femoral components (60-63), independent of age of the patient. The chapter on "Stem Fracture" (Chapter 7) also reflects the combined effects of improved stem design, stronger materials and improved femoral cementing in the elimination of this abrupt, severe complication.

Similarly, all these features of improved cementing and improved design aggregated to reduce sharply the generation of polymethylmethacrylate particles from loose femoral components (50) and thus reduce their contributing role to failure from and reoperations for the single, leading long-term complication, periprosthetic osteolysis. (See periprosthetic osteolysis in Chapter 11.)

Equally so, on the acetabular side for cemented components, substantive advances arose from our Laboratory as well. Charnley's innovative approaches to acetabular cementing, which stood as the world's standard in the 1960's, while revolutionary, had many disadvantages. His reamer required drilling a centralizing pilot hole through the medial wall of the acetabulum. This meant access into the pelvis for cement and all of the disadvantages thereof including access for intrapelvic infection, damage to pelvic structures, a pelvis mass of cement, difficulties in revision and loss of the ability to pressurize the cement. Thus, simply using hemispheric acetabular reamers on a power tool which I had developed for cup arthroplasty surgery overcame all these deficiencies.

Of course, the strengthening of the cement itself by porosity reduction applied here as well as on the femoral side. The two distinctive additional contributions to longevity of cemented acetabular components were a) pressurization of the cement (25) and b) the improvement in stress distribution in the adjacent bone as well as within the cement mantle itself by metal backing the component (64-71).

This was designed primarily to allow replacement of worn polyethylene liners, but also the metal backing strengthened the implant-cement interface. This concept was the forerunner of the Harris-Galante modular cementless acetabular component, which is now a world standard (See Chapter 9).

Our work on quantifying the fatigue effect of adding antibiotics to bone cement (72-74) provided important information about this critical material in its use both to prevent and to treat prosthetic local infection when cementing an acetabular component. All antibiotics weakened the cement and greater doses weakened it more.

Our extensive work on augmenting the available bone stock in the acetabular region for cementing in an acetabular component by adding femoral head autografts or femoral head allografts (75, 76, 77, 78, 79, 80, 81) played an important role in allowing further development of total hip replacement for dysplasia and developmental dislocation cases and other bone deficiency states.

In contrast with the behavior of the early cemented femoral components, loosening of a cemented acetabular component because of cement fatigue failure was uncommon except in instances of markedly compromised techniques. Rather, the common mechanism for acetabular component loosening was exactly the opposite, i.e., linear periprosthetic osteolysis leading to the resorption of the supporting bone by macrophage-induced osteoclastic activity secondary to particle disease at the bone-cement interface (50), not debonding or fatigue failure of the cement. Loosening of the cemented acetabular component was biological in origin, not mechanical, in most cases. The very important subject of periprosthetic osteolysis and the critical role of the Harris Orthopaedic Laboratory in it, from the first identification of this linear form of loosening of the cemented acetabular component (50) is detailed in the accompanying companion book on the life story of periprosthetic osteolysis. It relates the early observations and contributions to the understandings of the molecular biology involved throughout, to its elimination by the invention of the highly crosslinked electron beam polyethylene. It is entitled "Vanishing Bone: Conquering a Stealth Disease Caused by Total Hip Replacements" published by Oxford University Press.

Clearly, all these advances in cementing technique substantially improved the efficacy of fixation and thus the longevity of cemented fixation for total hip surgery. The dramatic nature of these results were quantified, by among others, through the Swedish National Total Hip Registry (60). In their 2000 publication the Swedish Registry reported a reduction of aseptic loosening (with or without osteolysis) of cemented total hip replacements, which had been the cause of 71% of revision operations. Aseptic loosening had been reduced to the low level of only 3% at 10 years. Moreover, they attributed this primarily to the effectiveness of improved surgical techniques.

A key step in quantifying the results of applying these techniques, as well as providing a method of self-improvement for the surgeons' immediate radiographic feedback, was our development of several radiographic criteria for the evaluation quality of the cementing as well as the appropriate placement of the implant. Our first rating scheme (69) defined "definite, possible and probable" criteria of a loose femoral component. Subsequently we developed specific radiographic criteria ("A,B,C and D") for the quality of the femoral cement application (70). These gradings were widely adopted and permitted not only the important self-assessment by the individual surgeon, but also valuable and important standards for the level of cementing done in the reports of success or failure of various series of cemented cases across the world.

Long-duration follow-up studies of cemented femoral stems done using good stem designs and good cementing techniques show several survival rates without loosening of a range between 80 and 90% at 26 years and 85-95% at 20 years (71).

## CEMENTLESS FIXATION

Driven by the exciting possibility of obviating the failures of fixation of total hip femoral components which had been fixed using bone cement plus the excitement of inventing a new and different form of bonding the components to the skeleton without cement and then driven even more strongly by the widely accepted concept that periprosthetic osteolysis was caused by submicron particulate bone cement, i.e., cement disease, the development of "cementless" or bone-in growth fixation moved rapidly forward. This rationale was further stimulated in the later years by recognition of both the causality and the increasing incidence of the late loosening of cemented acetabular components from linear periprosthetic osteolysis at the acetabular cement-bone interface. Cementless fixation aggressively overtook the field of total hip replacement, particularly in the United States.

As mentioned above, all of the earliest embodiments of total hip replacement surgery were cementless, including the 19th century adventures of Themistocles Gluck, the first 20th century version by Philip Wiles and later by his protege Ken McKee (1). Even more surprising, this list also includes Charnley, himself. His first acetabular and femoral components were cementless designs. But, in the 60's, after he introduced methyl methacrylate, cement fixation dominated.

Therefore, faced with the two major compelling motivations, failure of fixation and avoiding "cement disease", a multiplicity of cementless designs and approaches were investigated. These included acetabular and femoral implants which could be screwed into the bone, and others which were fixed to the femur by transfixation screws. But the central concept was that of a porous surface on the prosthesis into which host bone would grow, creating a biologic lock. The concept predicted that such a biologic response would self-renew and overcome the late failure of fixation resulting from excessive stress leading to fatigue failure of the cement fixation.

Thus, the key development focused on the characteristics of the porous layer. The central considerations were the material, the pore size, pore configuration, number and thickness of the layers and the acceptable amount of micromotion which would still allow bony ingrowth. This latter aspect became translated into defining and achieving a high degree of rigidity initially from the press fit surgery.

The early experimental success of the titanium fiber mesh of Rostoker and Galante (82) was followed by the development of beaded chrome cobalt implants and subsequently by plasma spray coatings. Animal experiments suggested that the most effective pore sizes ranged between about 100 to 500 microns and that at least two layers of porous material appeared to offer an advantage.

For femoral fixation, the promotion of fully coated, beaded chrome cobalt femoral components by Engh (83, 84) and others advanced that portion of the field, followed by an intense rivalry with proximally coated porous stems.

After many failures, extensive clinical studies established that with proper stem design and surgical techniques to create relatively rigid initial fixation at surgery, all the three types of coatings noted above were effective. An important additional feature, also noted above, was that having <u>circumferential</u> proximal coating was essential for cementless femoral stems. In the early iterations of the Harris-Galante femoral stem, the manufacturer was unable to apply the fiber mesh pads circumferentially. The gap between the pads (so-called patch porous coatings) provided a ready pathway for polyethylene debris to traverse down the femur, allowing periprosthetic osteolysis to develop early and extensively (85, 86). Correction of this manufacturing limitation resolved that specific problem.

While the Harris Orthopaedic Laboratory contributed importantly to development of many aspects of cementless fixation, the most valuable contribution was in the form of the optimization of the design of the cementless acetabular component, which is described in the section on cementless acetabular design in Chapter 9. In short, the ultimate concepts manifest in the Harris-Galante acetabular component (87) were a hemispheric metal shell with a fiber mesh porous coating, either press fit or fixed with screws and since 2000 used with a replaceable highly crosslinked polyethylene liner. These features provided not only the ultimate durable fixation without revisions for wear or lysis, but this concept provided additionally the optimum flexibility in choosing the bearing material, the size of the ball, resistance to dislocation, low wear and little or no lysis in the general population (88-90) and even as in those 50 and younger (91, 92).

Our unique, sophisticated micromotion experiments provided the most exact information defining the requirement for the degree of initial stability that would allow bony ingrowth. This canine experiment established that even in the face of continuous oscillation micromotion of 20 microns, bone would grow in and successfully lock the experimental porous-coated titanium cylinder solidly to the femur (93).

In terms of <u>fixation</u> of components, cementless techniques now dominate worldwide for both <u>femoral</u> and <u>acetabular</u> components. The one population subgroup in which cement fixation still produces superior results is the femoral fixation in hip fracture patients over age 75 (94).

Because of the interplay of so many advances, including crosslinked polyethylene and ceramic-on-ceramic articulations, improved fixation of both cemented and cementless components, and advanced designs and improved materials, an overview statement that reflects the integrated results of all these features is important. A seminal example is the report from the Australian Orthopaedic Association National Joint Replacement Registry 2016 which found that at 15 years in total hip replacements using crosslinked polyethylene articulations and contemporary designs with current surgical techniques, the revision rate for a combined criteria of lysis plus loosening was only 1.1% (95). This is a most striking improvement.

## SELECTED RELATED REFERENCES

1.  Hernigou P. Earliest times before hip arthroplasty: from John Rhea Barton to Themistocles Gluck. Int Orthop 37(11):2313-2318, 2013.
2.  Waugh W. John Charnley: The Man and the Hip. Springer-Verlag. London Berlin Heidelberg New York Paris Tokyo Hong Kong 139-140, 1990.
3.  Gruen TA, McNeice GM, Amstutz HC. "Modes of Failure" of cemented stem-type femoral components: a radiographic analysis of loosening. Clin Orthop Rel Res Jun;(141):17-27, 1979.
4.  Beckenbaugh RD, Istrup DM. Total hip arthroplasty. J Bone Joint Surg Am Apr;60(3):303-313, 1978.
5.  Warth LC, Callaghan JJ, Lu SS, Klaassen AL, Goetz DD, Johnston RC. Thirty-five year results after Charnley total hip arthroplasty in patients less than fifty years old. A concise follow-up of previous reports. J Bone Joint Surg Am Nov 5;96(21):1814-1819, 2014.
6.  Ibid. pp. 139-144.
7.  Burke DW, Gates EI, Harris WH. Centrifugation as a Method of Improving Tensile and Fatigue Properties of

Acrylic Bone Cement. J Bone Joint Surg 66-A:1265-1273, 1984.

8. Davies JP, Burke DW, O'Connor DO, Harris WH. Comparison of the Fatigue Characteristics of Centrifuged and Uncentrifuged Simplex P Bone Cement. J Orthop Res 5(3):366-71, 1987.

9. Davies JP, O'Connor DO, Burke DW, Jasty M, Harris WH. The Effect of Centrifugation on the Fatigue Life of Bone Cement in the Presence of Surface Irregularities. Clin Orthop Rel Res 229:156-161, 1988.

10. Davies JP, Jasty M, O'Connor DO, Burke DW, Harrigan TP, Harris WH. The Effect of Centrifuging Bone Cement. J Bone Joint Surg. Br. 71-B:39-42, 1989.

11. Davies JP, O'Connor DO, Burke DW, Greer JA., Harris WH. Comparison and Optimization of Three Centrifugation Systems for Reducing Porosity of Simplex P Bone Cement. J Arthroplasty 4:15-20, 1989.

12. Davies JP, O'Connor DO, Burke DW., Greer JA, Harris WH. Influence of Antibiotic Impregnation on the Fatigue Life of Simplex P and Palacos R Acrylic Bone Cements, with and without Centrifugation. J. Biomed. Mat. Res. 23:379-397, 1989.

13. Jasty M, Davies JP, O'Connor DO, Burke DW, Harrigan TP, Harris WH. Porosity of Various Preparations of Acrylic Bone Cements. Clin Orthop Rel Res 259:122-129, 1990.

14. James S, Jasty M, Davies J, Piehler H, Harris WH. A fractographic investigation of PMMA bone cement focusing on the relationship between porosity reduction and increased fatigue life. J. Biomedical Mat. Res. 26(3):651-662, 1992.

15. Davies JP, Harris WH. Comparison of diametral shrinkage of centrifuged and uncentrifuged Simplex-P bone cement. J. Applied Biomaterials 6:209-211, 1995.

16. Gates EI, Carter DR, Harris WH. Comparative Fatigue Behavior of Different Bone Cements. Clin Orthop Rel Res, 189:294-299, 1984.

17. Rey RM Jr., Paiement GD, McGann WM, Jasty M, Harrigan TP, Burke DW, Harris WH. A Study of Intrusion Characteristics of Low Viscosity Cement Simplex-P and Palacos Cements in a Bovine Cancellous Bone Model. Clin Orthop Rel Res 215:272-278, 1987.

18. Davies JP, O'Connor DO, Greer JA, Harris WH. Comparison of the mechanical properties of Simplex P, Zimmer Regular, LVC Bone Cements. J. Biomed. Mat. Res. 21:719-730, 1987.

19. Thanner J, Freij-Larsson C, Karrholm J, Malchau H, Wesslen B. Evaluation of Bone Loc. Chemical and mechanical properties and a randomized clinical study of 30 total hip arthroplasties. Acta Orthop Scand. 6:207-214, 1995.

20. Harris WH, McGann WA. Loosening of the Femoral Component after Use of the Medullary-Plug Cementing Technique. Follow-up Note with a Minimum Five-year Follow-up. J. Bone Joint Surg., 68-A:1064-1066, 1986.

21. McLaughlin JR, Harris WH. A Composite Plug for Occluding the Femoral Canal Prior to Cementing a Total Hip Femoral Component. Orthopaedic Review 23:344-356, 1994.

22. Harris WH, Davies JP. Modern Use of Modern Cement for Total Hip Replacement. Ortho Clinics of North America 19:581-589, 1988.

23. Oh I, Bourne RB, Harris WH. The Femoral Cement Compactor. An Improvement in Cementing Technique in Total Hip Replacement. J. Bone Joint Surg., 65-A:1335-1338, 1983.

24. Bourne RB, Oh I, Harris WH. Femoral Cement Pressurization During Total Hip Arthroplasty. The Role of Different Femoral Stems with Reference to Stem Size and Shape. Clin Orthop Rel Res 183:12-16, 1984.

25. Davies JP, Harris WH. In Vitro and In Vivo Studies of Pressurization of Femoral Cement in Total Hip Arthroplasty. J Arthroplasty 8:585-591, 1993.

26. Oh I, Carlson CE, Tomford WW, Harris WH. Improved Fixation of the Femoral Component after Total Hip Replacement Using a Methacrylate Intermedullary Plug. J. Bone Joint Surg., 60-A: 608-612, 1978.

27. Oh I, Harris WH. A Cement Fixation System for Total Hip Arthroplasty. Clin Orthop Rel Res 164:221-229, 1982.

28. Oh I, Merckx DB, Harris WH. Acetabular Cement Compactor: An Experimental Study of Pressurization of Cement in the Acetabulum in Total Hip Arthroplasty. Clin Orthop Rel Res 177:289-293, 1983.

29. Davis NJ, Jennings JJ, Harris WH. Induced Hypotensive Anesthesia for Total Hip Replacement. Clin. Orthop Rel Res, 101:93-98, 1974.

30. Harris WH, Pierce RW, Davis NJ, Jennings JJ, Fahmy N. Induced Hypotensive Anesthesia for Total Hip Replacement. In: The Hip., Proceedings of the Fifth Open Scientific Meeting of the Hip Society, Ed. WR Murray, St. Louis: CV Mosby Co.: 267-273, 1977.

31. Patel D Jr, Moellering RC, Thrasher K, Fahmy NR, Harris W.H. The Effect of Hypotensive Anesthesia on Cepalothin Concentrations in Bone and Muscle of Patients Undergoing Total Hip Replacement. J. Bone Joint Surg., 61-A:531-538, 1979.

32. Davies JP, Harris, WH. Strength of Cement-Metal Interfaces in Fatigue: Comparison of Smooth, Porous and Precoated Specimens. Clin. Materials 12:121-126,1993.

33. Davies JP, Harris WH. Tensile Bonding Strength of the Cement-Prosthesis Interface. Orthopedics 17:171-173, 1994.

34. Davies JP, Tse MK, Harris WH. In Vitro Evaluation of Bonding of the Cement-Metal Interface of a Total Hip Femoral Component Using Ultrasound. J Orthop Rel Res. 13:335-339, 1995.

35. Harris WH. Long Term Results of Cemented Femoral Stems with Roughened Precoated Surfaces. Clin. Orthop. 355:137-143, 1998.

36. Clohisy JC, Harris WH. Primary Hybrid Total Hip Replacement, Performed with Insertion of the Acetabular Component without Cement and a Precoat Femoral Component with Cement: An Average ten-year follow-up study. J Bone Joint Surg.81-A (2):247-255, 1999.

37. Dowd JE, Cha CW, Trakru S, Kim SY, Yang IH, Rubash HE. Failure of total hip arthroplasty with a precoated prosthesis. 4-to-11-year results. Clin Orthop Rel Res. 355:123-136, 1998.

38. Brown 3rd EC, Lachiewicz PF. Precoated femoral component in total hip arthroplasty. Results of 5- to 9-year followup. Clin Orthop Rel Res. 364:153-159, 1999.

39. Ahmed AM, Raab S, Miller JF. Metal/cement interface strength in cemented stem fixation. Clin Orthop Rel Res 2:105-118, 1989.

40. Raab S, Ahmed AM, Povan JW. Thin film PMMA precoating for improved implant bone cement fixation. J. Biomed Mater. Res 16 16(5):579-704, 1982.

41. 41. Caler WE, Carter DR, Harris WH. Technical Note. Techniques for Implementing an In Vivo Bone Strain Gage System. J. Biomechanics. 14:503-507, 1981.

42. Gates-Pelander E, Carter DR, Harris WH. Strain Controlled Fatigue of Acrylic Bone Cement. J. Biomed. Mat. Res., 16:647-657, 1982.

43. Gates EI, Carter DR, Harris WH. Tensile Fatigue Failure of Acrylic Bone Cement. J. Biomech. Eng., 105: 393-397, 1983.

44. Harris WH. Will Stress Shielding Limit the Longevity of Cemented Femoral Components of Total Hip Replacement? Clin. Orthop. Rel Res 274: 120-123, 1992.

45. O'Connor DO, Burke JW, Jasty M, Sedlacek RC, Harris WH. In Vitro Measurement of Strain in Bone Cement Surrounding the Femoral Component of Total Hip Replacements during Simulated Gait and Stair-Climbing. J Orthop Rel Res 14:769-777, 1996.

46. Sedlacek RC, O'Connor DO, Lozynsky AJ, Harris WH. Assessment of the symmetry of bone strains in the proximal femoral medial cortex under load in bilateral pairs of cadaver femurs. J. Arthrop. 12:698-694, 1997.

47. Harrington MA, O'Connor DO, Lozynsky AJ, Kovach I, Harris WH. Effects of Femoral Neck Length, Stem Size, and Body Weight on Strains in the Proximal Cement Mantle. J Bone Joint Surg 84-A: 573-579, 2002.

48. Jasty M, Maloney WJ, Bragdon CR, O'Connor DO, Haire T, Harris WH. The Initiation of Failure in Cemented Femoral Components of Hip Arthroplasties. J Bone Joint Surg 73-B (4):551-558, 1991.

49. Harrigan TP, Kareh J, O'Connor DO, Burke DW, Harris WH. A Finite Element Study of the Initiation of Failure of Fixation of Cemented Total Hip Components. J Orthop Rel Res. 10(1):134-144, 1992.

50. Schmalzried TP, Kwong LM, Jasty M, Sedlacek RC, Haire TC, O'Connor DO, Bragdon CR, Kabo JM, Malcolm AJ, Harris WH. The Mechanism of Loosening of Cemented Acetabular Components in Total Hip Arthroplasty. Clin Orthop Rel Res 274: 60-78, 1992.

51. Davies JP, Harris WH. The Effect of Addition of Methylene Blue on the Fatigue Strength of Simplex-P Bone Cement. J. Applied Biomaterials 3:81-85, 1992.

52. Harrigan TP, Harris WH. A Three-Dimensional Non-Linear Finite Element Study of the Effect of Cement-Prosthesis Debonding in Cemented Femoral Total Hip Components. J. Biomechanics 24:1047-1058, 1991.

53. Davies JP, Singer G, Harris WH. The Effect of a Thin Coating of Polymethlymethacrylate on the Torsional Fatigue Strength of the Cement-Metal Interface. J. Appl. Biomaterials 3(1):45-50, 1992.

54. James SP, Schmalzried TP, McGarry F.J, Harris WH. Extensive Porosity at the Cement-Femoral Prosthesis

Interface: A Preliminary Study. J. Biomed. Mat. Res. 27:71-78, 1993.

55. Jasty M, Jiranek W, Harris WH. Acrylic Fragmentation in Total Hip Replacements and Its Biological Consequences. Clin Orthop Rel Res 285:116-128, 1992.

56. Kawate K, Maloney WJ, Bragdon CR, Biggs SA, Jasty M, Harris WH. Importance of a Thin Cement Mantle: Autopsy Studies of Eight Hips. Clin Orthop Rel Res 355:70-76, 1998.

57. Harris WH. The Importance of the Thickness of the Femoral Cement Mantle. Hip International. 13(2):61-64, 2003.

58. Estok DM II, Orr TE, Harris WH. Factors affecting cement strains near the tip of a cemented femoral component. J Arthroplasty 12:40-48, 1997.

59. Estok DM II, Harris WH. A Stem Design Change to Reduce Peak Cement Strains at the Tip of Cemented Total Hip Arthroplasty. J Arthroplasty 15:5, 584-589, 2000.

60. Herberts P, Malchau H. Long-term Registration Has Improved the Quality of Hip Replacement: A Review of Swedish THR Registries Comparing 160,000 cases. Acta Orthop Scand. 71(2): 111-121, 2000.

61. Mulroy RD Jr, Harris WH. The Effect of Improving Cementing Techniques on Component Loosening in Total Hip Replacement. J Bone Joint Surg 72-B:757-760, 1990.

62. Smith SE II, Estok DM, Harris WH. 20-Year Experience with Cemented Primary and Conversion Total Hip Arthroplasty Using So-Called Second-Generation Cementing Techniques in Patients Aged 50 Years or Younger. J Arthroplasty 15:3, 263-273, 2000.

63. Harris WH. Improved Long-Term Results of Femoral Fixation in Revision Surgery Using Modern Cementing. Ch. 31, p. 289-294. In: Total Hip Revision Surgery. Galante JO, Rosenberg AG, Callaghan JJ Eds. Bristol-Myers-Squibb/Zimmer Orthopaedic Symposium Series. New York, Raven Press:1995.

64. Vasu R, Carter DR, Harris WH. Stress Distributions in the Acetabular Region. I. Before and After Total Joint Replacement. J. Biomech., 15:155-164, 1982.

65. Carter DR, Harris WH, Vasu R. Stress Distributions in the Acetabular Region–II. Effects of Cement Thickness and Metal Backing in the Total Hip Acetabular Component. J. Biomech. 15:165-170, 1982.

66. Vasu R, Carter DR, Harris WH. Evaluation of Bone Cement Failure Criteria with Applications to the Acetabular Region. J. Biomech. Eng., 105:332-337, 1983.

67. Harris WH Jr., White RE. Socket Fixation Using a Metal-Backed Acetabular Component for Total Hip Replacement. A Minimum Five-Year Follow-up. J. Bone Joint Surg. 64-A:745-748, 1982.

68. Harris WH, Penenberg BL. Further Follow-up on Socket Fixation using a Metal-Backed Acetabular Component for Total Hip Replacement. A Minimum Ten-year Follow-up Study. J. Bone Joint Surg. 69-A:1140-1143, 1987.

69. 69. Harris WH Jr, McCarthy JC, O'Neill DA. Loosening of the Femoral Component of Total Hip Replacement after Plugging the Femoral Canal. In: The Hip, Proceedings of the 10th Open Scientific Meeting of the Hip Society, St. Louis: CV Mosby Co., 228-238, 1982.

70. 70. Barrack RL Jr, Mulroy RD, Harris WH. Improved Cementing Techniques and Femoral Component Loosening in Young Patients with Hip Arthroplasty: a 12 Year Radiographic Review. J Bone Joint Surg. 74-B:385-389, 1992.

71. Bedard NA, Callaghan JJ, Stefi MD, Liu SS. Systematic review of literature of cemented femoral components: what is the durability at minimum 20 years follow up? Clin Orthop Rel Res. Feb;473(2):563-571, 2 915.

72. Harris WH Jr, White RE. Socket Fixation Using a Metal-Backed Acetabular Component for Total Hip Replacement. A Minimum Five-Year Follow-up. J. Bone Joint Surg., 64-A:745-748, 1982.

73. Gerber SD, Harris WH. Femoral Head Autografting to Augment Acetabular Deficiency in Patients Requiring Total Hip Replacement. J Bone Joint Surg., 68-A: 1241-1248, 1986.

74. Davies JP, O'Connor DO, Burke DW, Greer JA, Harris WH. Influence of Antibiotic Impregnation on the Fatigue Life of Simplex P and Palacos R Acrylic Bone Cements, with and without Centrifugation. J. Biomed. Mat. Res. 23:379-397, 1989.

75. Harris WH, Crothers OD. Autogenous Bone Grafting Using the Femoral Head to Correct Severe Acetabular Deficiency for Total Hip Replacement. In: The Hip, Proceedings of the Fourth Open Scientific Meeting of the Hip Society. Ed: CM Evarts, St. Louis: CV Mosby Co., pp. 161-185, 1976.

76. Harris WH, Crothers O, Oh I. Total Hip Replacement and Femoral-Head Bone-Grafting for Severe Acetabular Deficiency in Adults. J Bone Joint Surg 59-A:752-759, 1977.

77. Harris WH. Allografting in Total Hip Arthroplasty: In Adults with Severe Acetabular Deficiency Including a Surgical Technique for Bolting the Graft to the Illum. Clin. Orthop Rel Res 162:150-164, 1982.

78. Jasty M, Harris WH. Total Hip Reconstruction Using Frozen Femoral Head Allografts in Patients with Acetabular Bone Loss. Orthop. Clin. of N. Am. 18:291-299, 1987.

79. Maloney WJ, Jasty M, Burke DW, O'Connor DO, Zalenski EB, Bragdon CR, Harris WH. Biomechanical and Histologic Investigation of Cemented Total Hip Arthroplasties. A Study of Autopsy-Retrieved Femurs After In Vivo Cycling. Clin Orthop Rel Res 249:129-40, 1989.

80. Mulroy RD Jr, Harris WH. Failure of Acetabular Autogenous Grafts in Total Hip Arthroplasty. Increasing Incidence: A Follow-Up Note. J Bone Joint Surg. (Am) 72-A: 1536-1540, 1990.

81. Kwong LM, Jasty M, Harris WH. High failure rate of bulk femoral head allografts in total hip acetabular reconstructions at 10 years. J. Arthroplasty 8:341-346, 1993.

82. Miller RA, Galante JO, Rostoker W. A porous titanium endoprosthesis for replacement of lost segments of long bones. Surg Forum 26:506-508, 1975.

83. Engh CA, O'Connor DO, Jasty M, McGovern TF, Bobyn JD, Harris WH. Quantification of Implant Micromotion, Strain Shielding, and Bone Resorption with Porous-Coated Anatomic Medullary Locking Femoral Prostheses. Clin Orthop Rel Res 285: 13-29, 1992.

84. Maloney WJ, Sychterz C, Bragdon CR, McGovern T, Jasty M, Engh CA, Harris WH. The Otto Aufranc Award. Skeletal response to well fixed femoral components inserted with and without cement. Clin Orthop Rel Res 333:15-26, 1996.

85. Goetz DD, Smith EJ, Harris WH. The prevalence of femoral osteolysis associated with components inserted with and without cement in total hip replacements. J Bone Joint Surg. 76-A :(8)1121-1129, 1994.

86. Bobyn JD, Jacobs JJ, Tanzer M, Urban RM, Aribindi R, Sumner DR, Turner TM, Brooks CE. The susceptibility of smooth implant surfaces to peri implant fibrosis and migration of polyethylene wear debris. Clin Orthop Rel Res 311:21-39, 1995.

87. Clohisy JC, Harris WH. The Harris-Galante porous-coated acetabular component with screw fixation. An average ten-year follow-up study. J Bone Joint Surg Am. Jan;81(1):66-73, 1999.

88. Lachiewicz PF, Soileau ES. Highly Cross-linked Polyethylene Provides Decreased Osteolysis and Reoperation at Minimum 10-Year Follow-up. J Arthroplasty Sep;31(9):1959-1962, 2016.

89. Nebergall AK, Greene ME, Rubash H, Malchau H, Troelsen A, Rolfson O. Thirteen-Year Evaluation of Highly Cross-Linked Polyethylene Articulating with Either 28-mm or 36-mm Femoral Heads Using Radiostereometric Analysis and Computerized Tomography. J Arthroplasty Sep;31(9 Suppl):269-276, 2016.

90. Hanna SA, Somerville L, McCalden RW, Naudie DD, MacDonald SJ. Highly cross-linked polyethylene decreases the rate of revision of total hip arthroplasty compared with conventional polyethylene at 13 years' follow-up. Bone Joint J. Jan;98B(1):28-32, 2016.

91. Greiner JJ, Callaghan JJ, Bedard NA, Liu SS, Gao Y, Goetz DD. Fixation and Wear with Contemporary Acetabular Components and Cross-Linked Polyethylene at 10-Years In Patients Aged 50 and Under. J Arthroplasty Sep;30(9):1577-1585, 2015.

92. Stambough JB, Pashos G, Bohnenkamp FC, Maloney WJ, Martell JM, Clohisy JC. Long-Term Results of Total Hip Arthroplasty with 28-Millimeter Cobalt-Chromium Femoral Heads on Highly Cross-Linked Polyethylene in Patients 50 Years and Less. J. Arthroplasty Jan;31(1):162-167, 2016.

93. Jasty M, Bragdon CR, Burke D, O'Connor DO, Lowenstein J, Harris WH. In vivo skeletal responses to porous-surfaced implants subjected to small induced motions. J Bone Joint Surg. 79-A: 707-714, 1997.

94. Nyholm A, Palm H, Malchau H, Troelsen A, Gromov K. Lacking Evidence for Performance of Implants Used for Proximal Femoral Fractures – A Systematic Review. Injury (Impact Factor:2:14). DOI:10.1016/j.injury.2016.01.001.

95. Australian Orthopaedic Association National Joint Replacement Registry p. 137, 2016.

CHAPTER 9

---------------------------------

# IMPLANT DESIGN

## THE OPTIMUM ACETABULAR COMPONENT DESIGN

THE MULTIPLICITY OF CONFLICTING DEMANDS presented by attempts to replace the acetabular portion of the human hip with a total hip replacement confounded success for decades. This is true despite the fact that Charnley's decision to change from his original cementless efforts to the use of cement fixation and then to change from PTFE to polyethylene, which appeared for years to solve the initial high incidence of failure in metal- on-plastic total hip prostheses. Only much later did the field acknowledge the compelling problems of late loosening and periprosthetic osteolysis associated with conventional polyethylene. Moreover, the additional issues of limited range of motion, small head size and dislocation lurked in the background from the early days.

Progressively, painful experience focused attention on five cardinal criteria for a successful acetabular component for total hip replacement: 1) "permanent" fixation to the skeleton, 2) extremely low wear, 3) wide range of motion, 4) resistance against dislocation, and 5) durability of the locking mechanism of modular liners.

This history of the evolution of the optimization of the acetabular component replacing the hip is instructive. One early approach to hip replacement was without <u>any</u> acetabular component. The successful experiences of the hemi-arthroplasty in the management of femoral neck fractures encouraged some surgeons to treat osteoarthritis and avascular necrosis of the hip simply with hemi-arthroplasty implants, ignoring the acetabular reconstruction. A particularly revealing example was McKee's experience. Initially he did this type of reconstruction for total hip surgery until he subsequently was forced to add the metal acetabular component designed by Watson-Farrar. That specific metal acetabular component, made of chrome cobalt, was fixed to the skeleton by cement, following Charnley's adoption of bone cement for the fixation of all THR components.

Charnley's innovative studies of the coefficient of friction of native joints were startling and revealing, identifying the extremely low coefficient of animal and human joints. This led him to postulate that any successful joint replacement must be characterized by "low friction". His first version of this was a Teflon (or PTFE) surface replacement design replacing basically the cartilage of both the femoral head and acetabulum. (see Figure 9.1A). Both shells were "press fit". Teflon was chosen because of its remarkable "slipperiness" (low coefficient of friction), although still higher than native joints.

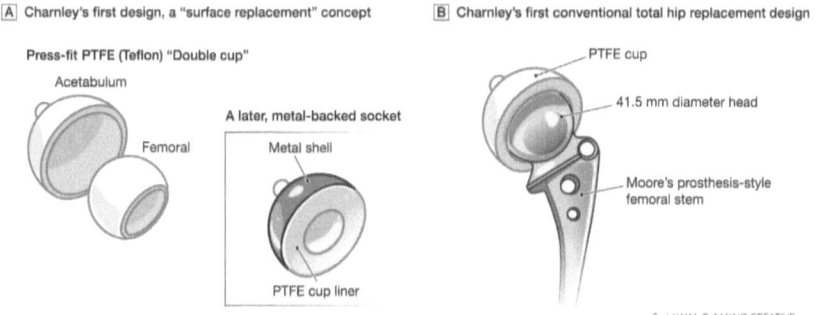

**Figure 9.1 A** Charnley's early total hip designs. Illustration A shows his first design, a "double cup". Later, Illustration B, he matched a "Teflon"socket with a 41.5mm diameter Moore type stem. Illustrations © Amino Creative. Printed with permission.

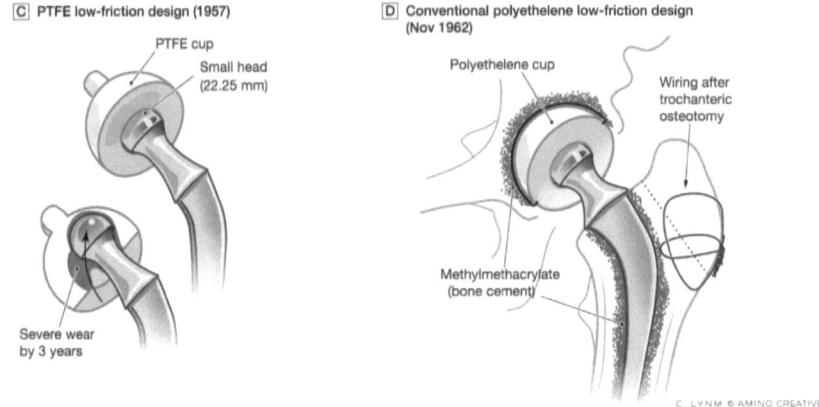

**Figure 9.1B Charnley's later total hip designs Illustration C shows the "Teflon"socket with the Charnley stem and the 22.5 mm head diameter. The lower figure illustrates severe wear at 3 years. Illustration D shows the polyethylene socket and shows both the bone cement and the trochanteric osteotomy.** Illustrations © Amino Creative. Printed with permission.

Changes in the <u>usage</u> of the term "press fit" warrant a word of explanation. Today, press fit design for fixation universally means "press fit with a bony ingrowth or bony ongrowth" onto or into the external surface, but in the earlier usage (19th century and 20th century up through the 1970's) the term was used without implying any bony fixation. Thus, when Charnley initially used "press fit" implants, these devices had no capacity for bony ingrowth or ongrowth.

While the transient relief of pain and range of motion were impressive in the initial attempts at Charnley's initial press fit surface replacement design, failure of fixation of the implants forced him to change. One of those changed designs actually presaged the subsequent success of today's cementless acetabular implants to the extent that he designed the acetabular component that consisted of a thin metal shell containing a plastic liner. See insert in Figure 9.1A, illustration A. However, the use of these two Teflon surface replacement, press fit components led to an evolution from his original thinking to combine a press fit Teflon acetabular component articulating against a femoral component modeled on the Moore hemiarthroplasty, that was also press fit into the femur. See Figure

9.1A, illustration B. This design, notably, employed a ball diameter of 41.5 mm, indicating that his ideal first concept of a total hip with a stemmed femoral component had a large femoral head. But failure of these early "press fit" devices ultimately led him to fix both components directly to bone using methyl methacrylate. See Figure 9.1B, illustration D.

Then, after being driven to recognize the remarkably rapid wear of his Teflon plastic, (see Figure 9.1B, illustration C.) he was forced eventually to make two dramatic changes, one on the femoral side and the other on the socket side. The first was to commit to progressively smaller ball diameters, because of the rapid wear of the Teflon. See Figure 9.1B, illustration C. He ultimately used a ball that was 22.5 mm in diameter. See Figures 9.1B, illustrations C and D. The second was to abandon Teflon completely and pivot to the use of conventional ultrahigh molecular weight polyethylene. See Figure 9.1B, illustration D.

In terms of wear, this latter combination of a small ball size articulating against the polyethylene was dramatically superior to the metal-on-Teflon plastic. But his concern about the the rapid wear rate of the Teflon material was so strong that even against the polyethylene, he persisted in advocating this very small head diameter throughout his entire career. See Figures 9.1B, illustration D.

At first, this combination of a small metal head articulating against polyethylene was remarkably successful, gradually superseding many of the alternative designs or techniques. It was not until his studies after eight or more years of use and then the later study at 12-15 years that Charnley reported three key adverse radiologic observations about his cemented sockets, even when using his polyethylene. These were 1) "cavitation", a form of balloon-like erosive periprosthetic osteolysis, 2) "demarcation", newly formed, linear radiolucency at the cement-bone interface and 3) even more striking, migration of the entire complex of the acetabular component plus its surrounding cement. These three alarming findings plus parallel observations made by others including our report of four cases of erosive periprosthetic osteolysis in the femur created widespread apprehension. That, in turn, generated a vigorous pursuit of the etiology of these strange forms of bone resorption. See Chapter 11.

Figure 9.2A, illustration A shows two hemispheric metal-backed designs of acetabular components for use with cement, the first one (the original modular acetabular cemented component) I introduced in 1971 and the latter modification as both cemented and as cementless acetabular component I introduced in 1983. Since modularity permitted any choice of liner, our highly crosslinked polyethylene was recommended once it became available.

As extensive studies supported the innovative hypothesis proposed by Willert and co-authors identifying the macrophage response to particulate bone cement as the underlying cause of cavitation, further confirmation studies ultimately led to Hungerford's arresting publication in 1987 which focused acutely on "Cement Disease".

As the data supporting Willert's hypothesis grew and the incidence of erosive bone resorption multiplied, massive efforts next arose to create an alternative or "cementless" method of fixation of the component to the skeleton, thus eliminating the use of bone cement. Ultimately, the successful form of this movement was "bony ingrowth or bony ongrowth" fixation. Use of this innovation was further reinforced with each successful report at longer durations of follow-up, also augmenting the increasing pressure for change which was resulting from the relentlessly increasing reports of "cavitation, demarcation, migration" at acetabular components and the loosening of femoral components from periprosthetic osteolysis in cemented THA.

But also many failures occurred among the early cementless designs and techniques at first. Despite this, the thrust to abandon bone cement grew exponentially. Thus, the key hurdle became how to define the optimum conditions for cementless components in terms of the details of the porous surface, design of the implants themselves, and the techniques of insertion. Ultimately the three most successful porous surfaces were the fiber metal titanium layer proposed by Rostoker and Galante, the beaded chrome cobalt layer advocated by Howmedica and the plasma spray porous surface supported by several manufacturers. Experimental and human studies gradually optimized critical features such as bead size, pore volume, depth of porous layer, etc.

But it was only much later that a very basic question concerning successful bony ingrowth was answered, namely what is the maximum amount of allowable micromotion which was compatible with successful bony ingrowth. Greater micromotion produced fibrous rather than bony fixation. We defined the answer to that critical issue in a complex mechanical-biologic experiment in the distal, lateral femoral condyle of dogs. Porous cylinders were implanted at that site which were continually oscillated for 8 hours a day along their long axis in different but known amounts of micromotion. At sacrifice after six weeks, the tissues in and around the porous layer were assessed histologically. From this experiment we established that the biology of bone ingrowth into the fiber mesh porous spaces in the intact dog femur readily occurred despite micromotions of up to 20 microns. If the range of micromotion exceeded that, fixation by bony ingrowth was lacking. Even if bone did grow into some of the fiber mesh in the face of micromotion greater than 20 micra, that bone was not in continuity with the host bone surrounding the implant.

Because this limitation had not yet been established during the early experiences in human use of cementless acetabular components, in addition to other negative design features of the early implants, these conditions led to fairly high rates of failure of cementless sockets in many of the early cementless designs. An example was the PCA cementless acetabular component, introduced in the early 1980's, using the chrome-cobalt beaded porous surface. It was of hemispherical design with two porous-covered pegs extending proximally from the superior-lateral portion of the component. These pegs were driven into recesses drilled into the innominate bone at the edge of the acetabular cavity. Because this mechanism did not sufficiently restrict micromotion, failure of fixation of this component to the pelvis was common. Callaghan et al. reported 23% of these acetabular implants had been revised for lysis or loosening by 15 years. In an important contrast, in that same study the corresponding analysis of the Harris-Galante cementless acetabular component (see

Figure 9.2B, illustration C) showed that at 15 years <u>none</u> were loose. Another alternate design used a cube shape in ceramic material, relying on the depth and multiple right angle corners to prevent rotation of the implant. Loosening rates with this design were also high.

Over time the essential steps in the development of the successful cementless socket were shown to be a) establishing rigid fixation of the porous metal shell, b) selection of an effective porous layer, c) use of a hemispherical design, d) developing the optimum material for the liner, and e) fixing the liner rigidly to the shell.

In practical terms for clinical usage, selection of the porous metal surface for each design of an implant was determined by two contradictory factors, the supporting experimental data versus the specific manufacturer involved, because each designer was beholden to the specific manufacturer who was willing to build components for him/her. Each manufacturer had access to and the ability to make only certain of the porous metal surfaces.

Conceptually, the hemisphere shape was best, based on both the anatomy of the human acetabulum and the need for a moderately thick plastic layer surrounding a spherical femoral head. But the definitive question was how to control micromotion in all directions, including tilt, migration and rotation.

In order to establish initial rigid fixation, in the early days that meant to me the mandatory use of screw fixation at surgery. Therefore, in the initial phase of the development of my ideas, both experimentally and subsequently clinically, the hemisphere design included outriggers through which screws could be placed. I specifically designed the three outriggers so that separate screws could be placed independently through the outriggers into the ilium, ischium and pubis. In our dog experiments the design worked beautifully. Because Howmedica was the company working with me on these designs, the porous surface on these hemispheres consisted of chrome cobalt beads.

Obviously the surgical exposure required to enable placement of the outriggers against these three different periacetabular surfaces was greater than that needed for placement of a simple hemisphere. Despite this disadvantage, the bony ingrowth in the dogs into those components was excellent and the same proved true in humans. While this design met all key criteria very well, the operation was difficult, requiring release of the iliopsoas tendon in some patients to obtain the exposure necessary to insert the screws. Since any widespread adoption of a hemispherical cementless acetabular component required that the operation be made easier, the obvious solution was to drive the screws directly through the shell and eliminate the outriggers.

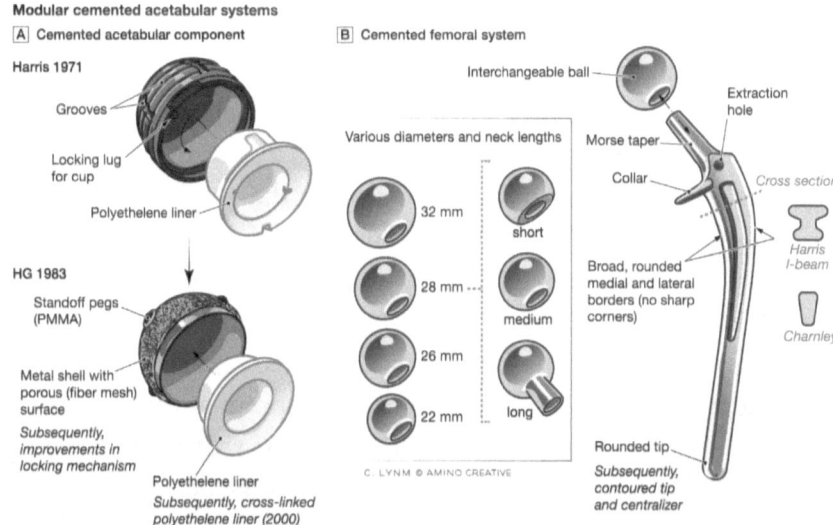

**Figure 9.2A Cemented total hip implants developed by William H. Harris and Jorge Galante.** Illustration © Amino Creative. Printed with permission.

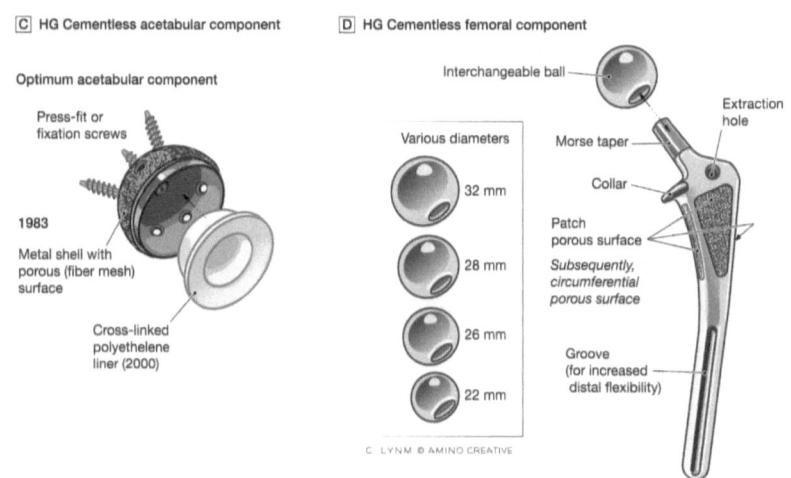

**Figure 9.2B Cementless total hip implants developed by William H. Harris and Jorge Galante.** Illustration © Amino Creative. Printed with Permission.

This was my next development. Plus one other. That other was a change in manufacturer, a transition from Howmedica, Inc. to Zimmer, Inc. This change resulted in a change in the porous layer. Titanium fiber mesh replaced the chrome cobalt beaded surface. See Figure 9.2A and B, illustrations A, C, and D.

Thus the final iteration for the acetabular shell for a cementless acetabular component was hemispheric, coated with porous fiber mesh and permitting screw fixation with the screws penetrating the hemisphere. See Figure 9.2B, illustration C. For over 34 years it has proved to be the ultimate design for "permanent" fixation to the skeleton of any cementless acetabular shell, outlasting all other acetabular component designs, either cemented or cementless.

A next important modification of this concept of the cementless acetabular component was the demonstration that when properly inserted with a tight press fit, the screws could be deleted.

The initial reaction to my concept of the ideal cementless acetabular component by other experts was surprising. Prominent learned and skilled total hip experts widely proclaimed that this concept was "a severe disaster about to happen". Despite these theoretical concerns by important and experienced innovators, it has succeeded extremely well and is unsurpassed for fixation to the skeleton.

Because at that time my concepts were being developed with Zimmer, Inc., another totally unexpected but greatly welcomed parallel set of ideas appeared when previously secret information revealed that Jorge Galante had also been working on a similar concept for a cementless acetabular component. Specifically this was also a hemispherical, fiber mesh porous-coated acetabular component transfixed with screws. See Figure 9.2B, illustration C.

When Zimmer proposed that we work together on the project, I was delighted because Jorge had, many times over, proven his remarkable ability as an innovator, showing great skill, outstanding insight and remarkable integrity. Thus, was the Harris-Galante acetabular component born. It is now the dominant worldwide concept in acetabular component design, adopted extensively by virtually all manufacturers.

Subsequently, one more major advance occurred. This general design concept was further greatly enhanced by adopting the trabecular metal porous coating and all the outstanding creative features which <u>that</u> advance allowed, including trabecular metal augments. The ultimate challenge for any acetabular reconstruction was the use of this concept in its application with trabecular metal coating and trabecular metal augments in extreme revision cases characterized by massive bone loss. The final step was the integration of these optimized features plus the use of a cage, leading to the successful management of extraordinarily massive bone loss, the cup-cage concept.

An important issue in making the concept of screw fixation of a hemispheric cementless acetabular successful was the wide range of the lengths and types of screws themselves. Because of the nature and anatomy of the innominate bone, having both cancellous and cortical screws was advantageous, and at remarkably differing lengths. In certain directions, the purchase of the screws was only in cancellous bone. In other directions, they might engage cortical bone. Frequently the available bone was quite thin, requiring very short screws and at other times screws up to 70 mm long would be required to obtain rigid purchase. And always great care was mandatory to avoid vascular or nerve damage from screws penetrating the bony confines in vulnerable regions. For the safe use of screws, substantial education of the hip surgeon was necessary, for none were previously experienced in avoiding these risks.

I innovated several techniques to aid in the safe fixation with the screws. A constant locus of excellent bone stock for the screws was that portion of the innominate bone just proximal to the sciatic notch. But obviously, placement blindly in the region of the sciatic notch was dangerous. These dangers could be virtually eliminated by passing the index finger <u>through</u> the sciatic notch, which instantly protected those vulnerable structures passing through it and also provided excellent tactile data for both the orientation of the drill hole for the screw and a safety mechanism if the drill bit or screw tip began to penetrate out of the bony confines.

It also proved to be of great value in difficult osteoporotic cases to use a long screw passing along that especially thickened region of the wing of the ileum extending toward the iliac crest. Fortunately that thick portion of the iliac wing led directly to a corresponding wider portion of the iliac crest. Thus, after the surgeon identified that thickened region of the iliac crest by palpation of the crest, the long screw could be directed blindly but safely along that thicker region to maximize the purchase of that screw in the iliac wing.

The outstanding, three-decade experience of excellent fixation plus two other features, namely modularity which permitted the optimization of the choice of material of the acetabular surface and the introduction of trabecula metal with augments (plus the cup-cage concept for the most severe cases of bone loss) have made this design the leading design in the world.

## Femoral Stem Design

Charnley, in introducing metal-on-polyethylene stemmed femoral components, chose to replicate a Moore hemiarthroplasty design. See Figure 9.1A, illustration B. McKee, earlier, had followed the design of the Thompson hemiarthroplasty. In both instances these choices, while appearing appropriate at the time, failed to optimize the design in terms of minimizing stress risers in the cement and in terms of reducing the risk of stem fracture if subsequent cement failure created a loose proximal portion of the stem with a fixed distal portion. In designing my first total hip replacement, the Harris 1 THR, I adopted these errors as well.

When the high incidence of cemented femoral stem loosening combined with the alarming but low-incidence problem of stem breakage forced recognition of these shortcomings, I introduced major advances in the form of the CAD design. It was named in reference to our use of the computer analysis as "computer-assisted design". A central factor was the broad medial border in contrast to the prior narrow border, to reduce the high stress on the medial cement. Also, all sharp corners were smoothed to decrease peak stress at those locations, especially as related to high rotatory stress in addition to the varus forces involved. Similarly the lateral borders were broadened to increase the resistance to stem fracture, in part because this stem continued to use the weaker cast material. The only way to protect the lateral region of the stem which was vulnerable to the initiation of fracture of the stem was to make it larger. That increased the resistance to fracture from rotatory strains as well.

An additional step was to incorporate the principles of the I-Beam construction by having both the medial and the lateral regions enlarged while reducing the mass of the metal in the central region of the stem. By so doing, and doing it in a way that both avoided sharp corners and also permitted extraction of the stem from the cement mantle if needed, further resistance to rotation was achieved as well. See Figure 9.2A, illustration B.

These features proved to be effective, with excellent, long-term results being obtained from this design but, because that design was still made of cast material, it was necessarily a very bulky stem.

Next in the progression of cemented femoral stem design from the Harris Orthopaedic Lab was the HD-2 design. It retained all the conceptual advantages of the CAD but, because it was fabricated from forged chrome cobalt, the bulk could be substantially reduced while affording even greater strength and ease of introduction and a more uniform and thicker cement mantle.

Along with these design features aimed at reducing peak cement strains and prolonging the life of the cemented fixation, a quite different but very important conceptual advantage also marked this new design. That was the innovative concept of having the stem available in 4 head diameters. Innovative, here, refers to the unique marriage of all these improved features of the stem design with different head diameters. See Figures 9.2A and B, illustrations B and D. This had never been done before, because previously designers of total hip replacement cemented femoral stem design had so strongly espoused one head diameter that they had insisted on their stem being available in only that head diameter. As is discussed in the chapter on "Systems", I felt that the HD-2 stem features were both sufficiently unique and sufficiently valuable, that if a surgeon was attracted to them, his or her use of the stem should not be blocked just because his or her preference was to use a head diameter other than 26 mm, which was my choice. All of these stems were of a monoblock design, i.e., without Morse taper.

Our next contribution in cemented stem design arose as an integral part of the "Total System", designed in 1982 and released in 1983. While the Total System integrated for the first time anywhere both cemented and cementless components for both the femur and the acetabulum into a single inclusive, unified system, I retained the central feature of the HD-2 femoral component in place. But I added one major change in the femoral components, the adoption of the Morse taper for interchangeability of both the diameter and the length of the femoral head components for each. This was a huge advance in terms of inventory while preserving maximum availability regarding head diameter, by lengths, and in those cases of revision surgery with an intact acetabular component, this flexibility in choosing the femoral head diameter permitted adjustability to the existing inside diameter of the retained acetabular component. See Figures 9.2A and B, illustrations B and D.

The further importance of this unique creation of an integrated system for both cemented and cementless components and both the femoral and the acetabular components is emphasized in the chapter dealing with "Systems". It was clear that this broad system integration of cemented and cementless components was highly propitious. This was particularly true during the era when hybrid components were so widely used. In the long run, this innovative concept of creating an integrated cemented plus cementless system became another world standard which arose from the Harris Orthopaedic Lab.

In contrast with the importance of the contribution which the Harris Orthopaedic Laboratory made to cementless acetabular design, in which our design and materials innovation solved all three key issues, namely, fixation, replaceability and wear resistance, our contributions to the cementless stem design were modest and in one instance, inadvertently negative.

When Jorge Galante and I joined forces to develop the Total System, we favored the fiber mesh that he and Rostoker developed. We designed initially a proximally-coated cylindrical stem, followed later by fully-coated cylindrical and a then proximally-coated anatomic implant.

The distinguishing feature of our first design was the fiber mesh porous coating. While the fiber mesh was standard in its characteristics, it was non-standard in distribution on the stem, meaning that at that time (1983) Zimmer was unable to fabricate it as a continuous circumferential layer over the proximal region of the stem. As a result, it was a "patch porous" layer, with the patches of the fiber mesh separated from each other by boundaries of smooth titanium. See Figure 9.2B, illustration D. Unknown at that time, this was a severe flaw.

The flaw lay in the pathway that these smooth boundaries provided for particulate polyethylene to use in accessing the "effective joint space", leading over time to produce lysis on the femur in high frequency.

In what was an inadvertent but well-designed experiment we compared incidence of femoral lysis in comparable patients after having either a cemented femoral stem or this "patch porous" cementless femoral stem. All patients had the same acetabular component, same head size, and the same conventional polyethylene from the same manufacturer. In just six years the lysis in this group of patients who had this cementless stem in the femur was 29% while among those in the cemented group was zero. If one assumes that the comparable acetabular circumstances in both groups produced similar loads of particulate polyethylene, these data indicate a marked difference in susceptibility to particle migration down the femur along the smooth titanium borders of the "patches" of fiber mesh and thus the pathway to produce lysis.

It was this clinical study plus the experimental work of Bobyn et al. that clearly proved the pathway hypothesis. Because these observations were unknown and unanticipated at the time, we had accepted this manufacturing limitation. These subsequent adverse observations contributed widely to a better understanding of periprosthetic osteolysis, the concept of the effective joint space and the optimization of cementless femoral component design. This manufacturing flaw was overcome, and, once corrected, these cementless stems, as well as the subsequent iterations including the more anatomic design, proved effective.

Many other innovators developed many different cementless femoral stems extensively, including the effective stem design of Wagner. These designs plus several modular designs became of great value in selected primary as well as revision surgery.

## SELECTED RELATED REFERENCES

McKee GK, Watson-Farrar J. Replacement of Arthritic Hips by the McKee-Farrar Prosthesis. J Bone Joint Surg (Br) May 48(2):245-259, 1966.

Galante J, Rostoker W, Lueck R, Ray RD. Sintered fiber metal composites as a basis for attachment of implants to bone. J Bone Joint Surg (AM) Jan; 53(1):101-114. PMID:5540151, 1971.

Galante J, Rostoker W. Fiber metal composites in the fixation of skeletal prosthesis. J Biomed Mater Res 7(3):43-61, 1973.

Griss P, Von Adrian-Werberg H, Krempien B, Reipa S, Lauterbach HJ, Hartung HJ. Morphological and biomechanical aspects of AI203 ceramic joint replacement. Experimental results and design consideration for human endoprostheses. J. Biomed. Mater Res. July;9(4)177-188, 1975.

Harris WH, Schiller AL, Scholler J-M, Freiberg RA, Scott R. Extensive Localized Bone Resorption in the Femur Following Total Hip Replacement. J. Bone Joint Surg., 58-A: 612-618, 1976.

Willert HG. Reactions of the Articular Capsule to Wear Products of Artificial Joint Prostheses. J Biomed Mater Res: 11(2) 157-164, 1977.

Griffith MJ, Seidenstein MK, Williams D, Charnley J. Eight year results of Charnley arthroplasties of the hip with special reference to the behavior of cement. Clin Orthop Rel Res Nov.-Dec. (137):24-36, 1978.

Charnley J. Low Friction Arthroplasty of the Hip: Theory and Practice. Springer-Verlag Berlin Heidelberg New York: 75-83, 1979.

Hedley AK, Clarke IC, Kozinn SC, Coster J, Gruen T, Amstutz HC. Porous ingrowth fixation of the femoral component in a canine surface replacement of the hip. Clin Orthop Rel Res. March;(163):300-311, 1982.

Harris WH Jr, White RE, McCarthy JC, Walker PS, Weinberg EH. Bony Ingrowth Fixation of the Acetabular Component in Canine Hip Joint Arthroplasty. Clin Orthop Rel Res., 176:7-11, 1983.

Harris WH, Jasty M. Bone Ingrowth into Porous Coated Canine Acetabular Replacements: The Effect of Pore Size, Apposition, and Dislocation. In: The Hip, Proceedings of the 13th Open Scientific Meeting of the Hip Society, DV Mosby Co., Ed. RH Fitzgerald, St. Louis: 214-234, 1985.

Turner TM, Sumner DR, Urban RM, Rivero DP, Galante JO. A comparative study of porous coatings in a weight-bearing total hip-arthroplasty model. J Bone Joint Surg (AM) Dec;68(9):1396-1409, 1986.

Jones LC, Hungerford DS. Cement Disease. Clin Orthop Rel Res. Dec. (225):192-206, 1987.

Waugh W. John Charnley: The Man and the Hip. Springer-Verlag, London Limited, pp. 106, 107, 124, 125, 139-151,1990.

Willert HG, Bertram H, Buchhorn GH. Osteolysis in Alloarthroplasty of the Hip. Clin Orthop Rel Res 258: 108-120, 1990.

Wasielewski RC, Cooperstein LA, Kruger MP, Rubash HE. Acetabular anatomy and the transacetabular fixation of screws in total hip arthroplasty. J. Bone Joint Surg (AM) April;72(4):501-508, 1990.

Winter M, Griss P, Scheller G, Moser T. Ten-to-14-year results of a ceramic hip prosthesis. Clin Orthop Rel Res. Sept;(282):73-80, 1992.

Goetz DD, Smith EJ, Harris WH. The prevalence of femoral osteolysis associated with components inserted with and without cement in total hip replacements. J Bone Joint Surg. 76-A :(8)1121-1129, 1994.

Harris WH. The Problem is Osteolysis. Clin Orthop Rel Res. Feb(311);46-53 Review. PMID 7634590, 1995.

Bobyn JD, Jacobs JJ, Tanzer M, Urban RM, Aribindi R, Sumner DR, Turner TM, Brooks CE. The susceptibility of smooth implant surfaces to periimplant fibrosis and migration of polyethylene wear debris. Clin Orthop Rel Res. Feb;(311):21-29, 1995.

Willert HG, Semlitsch M. Tissue Reactions to Plastic and Metallic Wear Products of Joint Endoprostheses. Clin Orthop Rel Res. 333: 4-14, 1996.

Jasty M, Bragdon CR, Burke D, O'Connor DO, Lowenstein J, Harris WH. In vivo skeletal responses to porous-surfaced implants subjected to small induced motions. J Bone Joint Surg. 79-A: 707-714, 1997.

Callaghan JJ, Savory CG, O'Rourke MR, Johnston RC. Are all cementless acetabular components created equal? J Arthroplasty Jun;19(4 Suppl 1):95-98, 2004.

Willert HG, Buckhorn GH, Fayyazi A, Flury R, Windler M, Lohmann CH. Metal-on-Metal Bearings and Hypersensitivity in Patients with Artificial Hip Joints. A Clinical and Histomorphological Study. J Bone Joint Surg. 87(1):28-36, 2005.

Anseth S, Pulido PA, Adelson WS, Patil SP, Sandwell JC, Colwell CW Jr. Fifteen-Year to Twenty-Year Results of Cementless Harris-Galante Porous Femoral and Harris-Galante Porous I and II Acetabular Components. J Arthroplasty;45(5): 687-691, 2010.

Joglekar SB, Rose PS, Lewallen DG, Sim FH. Tantalum acetabular cups provide secure fixation in THA after pelvic irradiation at minimum 5-year follow-up. Clin Orthop Rel Res. Nov;470(11):3041-3047, 2012.

Long WJ, Noiseux NO, Mabry TM, Hanssen AD, Lewallen DG. Uncemented Porous Tantalum Acetabular Components: Early

Follow-Up and Failures in 599 Revision Total Hip Arthroplasties, Iowa Orthop J. 35:108-113, 2015.

Whitehouse MR, Masri BA, Duncan CP, Garbuz DS. Continued good results with modular trabecular metal augments for acetabular defects in hip arthroplasty at 7 to 11 years. Clin Orthop Rel Res;473(2):521-527, 2015.

Amanatullah DF, Howard JL, Siman H, Trousdale RT, Mabry TM, Berry DJ. Revision total hip arthroplasty in patients with extensive proximal femoral bone loss using a fluted tapered modular femoral component. Bone Joint J Mar;97-B(3):312-317, 2015.

Makinen TJ, Kuzyk P, Safir OA, Backstein D, Gross AE. Roles of Cages in Revision Arthroplasty of the Acetabulum. J. Bone Joint Surg (AM) Feb 3:98(3):233-242. doi:10.2106/JBJS.O.00143, 2016.

Amenabar T, Rahman WA, Hetaimish BM, Kuzyk PR, Safir OA, Gross AE. Promising Mid-term Results with a Cup-Cage Construct for Large Acetabular Defects and Pelvic Discontinuity. Clin Orthop Rel Res. Feb;472(2):408-414, 2016.

# CHAPTER 10

## THE SPECIAL PROBLEM OF DEVELOPMENTAL TOTAL DISLOCATION OF THE HIP

FROM TIME IMMEMORIAL, DEVELOPMENTAL DYSPLASIA of the hip (DDH) has represented a severe challenge to those patients afflicted with this malady, as well as to those surgeons attempting to ameliorate it. The range of abnormality varies from slight underdevelopment of the acetabulum with relatively normal development of the femur to those patients in whom the femoral head has never resided within the acetabular recess, i.e., those with total developmental dislocation of the hip, and to those who have severe deformity of the femur.

The very strong gender association is obvious with 95% of the patients being female and just 5% male. If the condition is unilateral, often the disparity between the leg length of the disabled limb and the normal limb is marked, contributing an additional factor to severe limping. Equally so, those patients with severe developmental dysplasia bilaterally also represent very special clinical compromises because of the severe waddling gait which can result from the bilateral involvement.

Prior to total hip replacement surgery, the limited variety of operative treatments available, specifically fusion, osteotomy, and cup arthroplasty, left very much to be desired. Fusion obviously could not be applied bilaterally. Cup arthroplasty and osteotomy could not improve the leg length discrepancy in unilateral cases. Moreover, the overall results of any of these operations were often quite limited. The best results in this group of relatively unsatisfactory options occurred in those patients who were candidates for and had successfully executed complex femoral or acetabular osteotomies, or both.

As was true for so many other hip conditions, total hip replacement represented a marvelous, near miraculous, dramatically new form of surgical correction. Charnley (1) imaginatively applied total hip replacements to many patients with mild and even marked dysplasia and, indeed, our experience with these cases gave remarkable results (2, 3, 4). But he drew the line in terms of those patients who had the total dislocation of the femoral head out of the hip socket. It was his feeling that total hip replacement could not be done in these patients (1).

One of the most surprising clinical aspects of this condition is that although many women with bilateral DDH and even with bilateral total dislocation from DDH suffered substantially in terms of their gait, they experienced little or no pain throughout adolescence, early adulthood and often even into late adult life. In fact, it is a fairly easy diagnosis to make from casual observations, simply watching an older woman with bilateral total dislocation exhibit the waddling gait so characteristic of this specific problem but clearly walking free of pain.

When I was presented with a patient with painful bilateral total dislocation from DDH during the early days of total hip replacement surgery in the United States in the early '70s, it seemed to me to be a compelling and intriguing challenge. Why not question Charnley's dictum (1) that those unfortunate patients, mostly women, with total dislocation should not have total hip replacement surgery.

In initially assessing the difficulties involved, Charnley's caution appeared well-founded. But several features were clear. First, one would have to have extremely small implants (5), both acetabular and femoral components, because of the severe hypoplasia that occurred in both skeletal areas. Conceptually, the question of the marked femoral anteversion of the femoral head and neck lent itself readily to correction with a cemented total hip replacement because the use of cement would allow the femoral component to be positioned in whatever degree of femoral anteversion the surgeon desired. The cement technique would accommodate that change in position. Still, the hypoplasia was a severe problem and none of the existing total hip replacement implants were compatible with the anatomic constraints, on the femoral or the acetabular side.

In the practice of medicine at that time, during the decades prior to three-dimensional imaging, it was not possible to determine either the volume of the acetabular recess or the volume of the medullary canal in the femur by radiographic techniques alone, as it is now. I elected to solve that problem surgically, by doing a preliminary operation to obtain direct measurements of these factors in the operating room. Such a preliminary operation was fully justified separately by the crucial need to be able to advance the femur substantially distally in order to bring the femoral head approximately into an area where it could articulate with a reconstructed acetabular recess. That, I felt, would require a release of the iliopsoas tendon and excision of the hip capsule, followed by postoperative traction. The measurements of the dimensions of the acetabular recess and the dimensions of the femoral head along with determination of the approximate size of the medullary canal could be done at the same time, after resection of the femoral head.

The femoral head, once amputated, could also serve as a bone graft to be fixed subsequently during the definitive operation against the wing of the ileum at the superior level of the acetabulum, augmenting the available bone stock to receive the cemented acetabular component.

That raised another interesting dilemma concerning the amputated femoral head. The question was what to do with the femoral head in the interim between this primary operation and the definitive operation. Again, this was at a time prior to the introduction of deep-freezing refrigerators for bone banks. It seemed to me, at that time and under those circumstances, that the thing to do was simply store it in the body. It would then be immediately available for bone grafting when the second, definitive operation was to occur.

Our experience with femoral head allografts and femoral head autografts in augmenting deficient acetabulae, primarily in other patients with developmental dysplasia of the hip but of a less severe degree, was substantial (6-16).

Armed with this plan, we undertook the world's first attempt to do a total hip replacement in an adult patient with total dislocation of the hip from DDH. And her other hip, also, had total dislocation of the hip from DDH. My patient, who also had the complicating disease of Turner's syndrome, was just 4 feet 8 inches tall, so she would have been small to begin with because of her Turner's syndrome, in addition to her underdevelopment specifically due to the developmental dislocation of both hips.

While we recognized that the anatomy would be severely distorted by this total dislocation of the hip, my concept of the full magnitude of the distortion of the anatomy would prove to be a serious underestimate. Specifically, because of the gross anatomic distortions caused by the total dislocation of the hip, the displacement of the profunda femoris artery was so severe that while I was doing the excision of the hourglass-shaped, distorted hip capsule, I inadvertently divided the profundus femoral artery. After controlling the alarming bleeding and obtaining reassurance from a vascular surgeon colleague that the collateral circulation in the limb would probably be adequate to keep the limb alive, we proceeded with the plan.

Fortunately, another completely surprising observation of that remarkable first operation was the fact that the limb had far more flexibility at the hip joint than I anticipated. In fact, after excision of the capsule and the release of the iliopsoas, I was able to bring the femoral head substantially distally toward the acetabular recess. It was not lax enough to reduce the hip fully but once the hip capsule had been excised and the psoas was released, it was nearly possible to "reduce" the hip.

Her postoperative course had one adverse effect as well. She developed myocardial ischemia. This meant that we could not proceed with the second surgery, the actual reconstruction of the hip, until at least three months of recovery from the myocardial ischemia. Since it would be prohibitively expensive to keep her in bed in skeletal traction for that additional three months just to pull the femur further distally than we could at surgery, that portion of the plan had to be abandoned. But we had already discovered the unexpected flexibility of the abductor musculature and had nearly been able to reduce the hip during this preliminary operation. So we felt it was probable that we could safely proceed with phase 2, the total hip in construction, once her myocardial ischemia had resolved, without using traction, and still reduce the femoral ball into the newly constructed socket.

The second operation initially proceeded just as planned. I recovered the amputated femoral head and bolted it to the innominate bone to augment the available bone for the acetabular reconstruction. The tiny, custom acetabular component fit perfectly into the reconstructed acetabular recess which I carved into the combined new bone stock, consisting of the existing bone available in her dysplastic acetabulum plus the grafted femoral head. The tiny (19 mm) femoral head could be reduced completely into the freshly cemented acetabular component and the greater trochanter (still attached to the abductors) could reach far enough distally to be wired to the lateral femoral cortex after I had positioned the new hip joint into maximum abduction.

The rest of the operation proceeded as planned, with, again, one exception. The femoral component, although being substantially straighter than anything that had ever been made before, still had too much of a curve in it for her small femur. As a result, the tip of the femoral component perforated the medial cortex of the proximal femur. Fortunately perforation of the medial cortex of the femur by the stem of a cemented femoral component carried much less risk of loosening than the same defect on the lateral cortex, with its severe varus position. In the long run this perforation caused no harm.

She was kept in balanced suspension in wide abduction for three weeks to allow the greater trochanter to begin to heal. Although the abductors were on maximum stretch at the end of the procedure, it was subsequently possible to gradually coax additional length from them slowly once she was being ambulated. Her limb gradually came out of the position of maximum abduction.

Otherwise her postoperative course was completely benign. The femoral head graft healed. The trochanter united to the femur. She progressively resumed full activities and became fully mobile, without pain and without support. Then we proceeded to incorporate the lessons of the first hip into the implant designs and surgical planning for the second hip.

I had learned from this first case that we could probably achieve this remarkable reduction of a completely dislocated, high-riding hip as a one-stage operation. I now also knew how severely distorted the anatomy would be. And I redesigned an improved femoral component. Her other hip operation progressed as planned, without incident. Her recovery was free of complications and totally successful. She made a full recovery from the second total hip operation also, becoming completely ambulatory without support or pain. It was a great success. She was delighted.

Thus, this initial experience led to a number of important observations that applied not only to those with total dislocations of the hip from DDH but to the far larger cohort, those with incomplete dislocations or dysplasia. An important subsequent major observation in dealing with additional patients with total hip dislocation from DDH was the realization that if a patient with this problem had the unusual but critical feature of a very stiff hip, the circumstances were entirely different. Although uncommon, such stiffness developed from progressive arthritic changes at the false joint. The critical aspect of this stiffness resulted from the lack of stretching of the muscle and therefore a marked lack of flexibility of the abductor muscles. They became shortened and rigid. This made it impossible to displace the femur distally to effect the reduction into the reconstructed socket. This made the completion of the operation – even using all of our specialized techniques and implants – impossible.

Because this specific observation had been totally unanticipated, its full impact was appreciated for the first time only in the operating room in such a case after the femoral head had been amputated and the commitment had been made to do a total hip replacement.

Faced with that dilemma several times, I developed two solutions, both successful. Faced with that stark problem, I was forced to improvise alternatives necessary to lengthen the key muscles surrounding the hip, predominantly the abductors. Of the two techniques for meeting this challenge, the first I created was to do a Z plasty in the abductors. I had no idea how well the abductors would function after a Z plasty but the patient who had the very first Z plasty rewarded me with a 25 follow-up visit. After 25 years of use, the Z plasty was quite successful, especially since after 25 years she walked without support and with a nearly normal gait. She also had a negative Trendelenburg test.

The alternative method I devised was to free the entire abductor muscle mass completely from the wing of the ilium all the way up to the iliac crest and then advance the abductor mass distally including its neurovascular bundles until the greater trochanter could reach the lateral femoral cortex with the limb in maximum abduction. But, this technique has the major disadvantage that in order to allow the abductors to heal back to the wing of the ilium in their new distal position, it is necessary to rest the patient in a plaster spica with both legs in wide abduction for six weeks.

Still, by using the inherent flexibility of the surrounding muscles in most cases of total dislocation of the hip from DDH, a complete capsulectomy and release of the iliopsoas or, if necessary, by using a Z plasty or by advancing the abductors, it became possible to address virtually any degree of severity of total dislocation from developmental dysplasia as a single operation.

Shortening the femur often became necessary to complete the reduction of the reconstructed hip joint, particularly if reducing the hip were likely to stretch the sciatic nerve and produce paralysis. If the operation was done by exposing the hip through a trochanteric osteotomy, the femur would be shortened by excising the necessary amount of femur and advancing the trochanter to the lateral surface of the remaining proximal part of the femur (4). A preferable technique was subsequently developed by Paavilainen, et al. (17) in which the trochanter was not osteotomized at all but rather the necessary shortening of the femur was accomplished by a <u>subtrochanteric</u> excision osteotomy. In that technique the proximal portion of the femur including the trochanter remained intact and was reamed and shaped to receive a femoral cementless component. After executing an excision subtrochanteric osteotomy, the femoral component is inserted into the proximal portion of the femur and extended into the medullary cavity of the shortened distal part of the femur. This technique became, by far, the preferable technique, particularly so using cementless femoral components.

All of these experiences in initiating the first total hip for a fully dislocated hip secondary to developmental dysplasia led to many important general understandings that contributed to other unusual problems of total hip reconstructions not directly related to DDH (18). For example, I learned a lot about what was necessary for some very difficult acetabular reconstructions not related to DDH, as well as what was required for some very difficult unrelated femoral reconstructions, and what might be required for optimum restoration of abductor function, or even what was required for preservation of distorted vasculature around the hip joint. In addition, I learned what limitations existed in terms of lengthening the leg without producing paralysis of the sciatic nerve.

The issue of effecting a major leg length increase without producing a sciatic palsy is complex. It does not come with a single definitive statement of a given length which can be or should not be exceeded. Rather, the individual conditions of the soft tissues, the overall height of the patient, the overall discrepancy between the original position of the femur and the subsequent position of the femur, and the amount of shortening of the femur all play a role. As a general rule, however, if the planning for a lengthening operation appears to increase the limb length of the overall reconstruction by more than a centimeter and a half, extra thought should be applied to the design and to the execution of this decision. Otherwise damage to the sciatic nerve could well occur. We devised several techniques for measuring and correcting leg lengths (19, 20).

The whole program of femoral head autografts and femoral head allografts to augment the acetabular bone stock ultimately prompted us to develop our own bone bank, prior to the more general creation of bone banks. An instructive case occurred in a patient who needed a standard total hip replacement on the left side but who also had developmental dysplasia with major acetabular deficiency on the right side. This led us, for the first time, to do the left hip initially so we could retrieve the left femoral head, freeze it in the bone bank until the right hip surgery was scheduled, and then use that as an autograft on the right side to augment the deficient acetabular bone stock.

This work substantially increased the number of patients who could undergo successful total hip replacement, despite major or even massive deficiency of the development of the acetabulum. From this experience I was then able to do long-duration follow-up studies to determine how these grafts functioned, studies of what was required to get them to unite, and other studies on what happened if acetabular fixations subsequently failed.

Just as I was the first to publish on femoral head autographs and the first to publish on femoral head allografts, I was also the first to publish on the duration before failures occurred of these large bulk cortico-cancellous grafts (14). Equally so our publications were the first to describe that, across major amounts of the grafted material, these grafts became revascularized. Moreover, many became sufficiently revascularized that if a revision operation became necessary, the vascularity was sufficient to permit bony ingrowth to progress into the cementless acetabular component from large areas of the revascularized grafted material (21, 22).

This effort in exploring the opportunities available in patients with developmental dysplasia led to the development of yet another concept for total hip replacement reconstructions, called the "high hip center" (23-28). In a number of patients with dysplasia, the acetabulum is not only hypoplastic, it is elliptical and eccentric. Despite this, almost always the total mass of available innominate bone for acetabular reconstruction exists at the site where the acetabulum has <u>actually</u> developed, independent of the site where it should have developed. With increasing experience, and particularly once cementless acetabular components became available, it became clear that small cementless acetabular components could play an important role in certain of these cases. Using a small femoral head, generally 22 mm. in diameter, and a 5 mm. thickness of polyethylene, many of these cases could be solved without acetabular grafting, simply by using the smaller cementless acetabular component. Often, however, this meant that an ultimate center of axis of rotation of the total joint reconstruction was substantially proximal to the normal location for the hip center. And yet, surprisingly and against the common beliefs, these reconstructions functioned extremely well. It was this line of reasoning that led to my concept of a "high hip center". In many instances the optimum solution was to use a hip center that was higher than normal, despite theoretical and mechanical disadvantages. Those theoretical disadvantages were not manifest in prohibitive loosening rates or in prohibitive wear rates of total hip replacements extant at that time, even using conventional polyethylene.

The next outstanding developments in the management of patients with developmental dysplasia of the hip were the addition of our cementless modular acetabular component (29-32) and the introduction of our highly crosslinked polyethylene (33-39). The singular distinguishing feature of the hemispherical cementless acetabular component, designed in conjunction with Jorge Galante, was the extraordinarily successful long-term fixation rate. While the Harris-Galante acetabular component in its original design had two shortcomings, namely poor fixation of the polyethylene to the shell and the use of conventional polyethylene, the fixation of the shell to the skeleton afforded by this design was truly remarkable. The long-term fixation rates are unexcelled by any other acetabular reconstruction of any type. For example, in a 17.2 year follow-up of 79 hips using the Harris-Galante acetabular component (with conventional polyethylene) in patients with a mean age of 54 years at the time of surgery, the remarkable figure was that only two acetabular components had been revised for loosening (40).

Moreover, both of the two original disadvantages of the initial design have been completely overcome. The first was overcome by developing better methods of fixing the polyethylene liner to the shell and the second was overcome by the development of highly crosslinked polyethylene. This cementless acetabular concept was subsequently enhanced greatly for revision cases by using trabecular metal as the shell itself or by using the trabecular metal as the porous coating and/or as augments (41-43).

The development of highly crosslinked polyethylene is the subject of Chapter 11 but clearly, in terms of patients with developmental dysplasia, as with all others, it provided an extraordinary advance in the longevity of the implant because of the low wear ratio and thus in the elimination of periprosthetic osteolysis.

Today, around the world, thousands upon thousands of patients with developmental dysplasia benefit from the central features of these unique developments of total hip replacement created specifically for these challenges, including those with total dislocation of the hip on a developmental basis. These patients, because they have lived throughout their lifetime prior to their total hip reconstruction surgery with major limitations of the hip, are among the most grateful of all patients benefiting from total hip replacement reconstructions.

## SELECTED RELATED REFERENCES

1. Charnley J, Feagin JA. Low Friction Arthroplasty in Congenital Subluxation of the Hip. Clin Orthop Rel Res. March-April (91):98-113, 1973.
2. Harris WH. Total Hip Replacement for Congenital Dysplasia of the Hip: Technique. In: The Hip, Proceedings of the Second Open Scientific Session of the Hip Society, ed. WH Harris, St. Louis: CV Mosby Co., 251-265, 1974.
3. Harris WH, Crothers OD. Autogenous Bone Grafting Using the Femoral Head to Correct Severe Acetabular Deficiency

for Total Hip Replacement. In: The Hip, Proceedings of the Fourth Open Scientific Meeting of the Hip Society. Ed: C.M. Evarts, St. Louis: CV Mosby Co., 161-185, 1976.

4. Harris WH, Crothers O, Oh I. Total Hip Replacement and Femoral-Head Bone-Grafting for Severe Acetabular Deficiency in Adults. J. Bone Joint Surg., 59-A:752-759, 1977.

5. Harris WH. Total Hip Replacement for Osteoarthritis Secondary to Congenital Dysplasia or Congenital Dislocation of the Hip. International Orthopaedics (SICOT) 2:127-138, 1978.

6. Harris WH. Allografting in Total Hip Arthroplasty: In Adults with Severe Acetabular Deficiency Including a Surgical Technique for Bolting the Graft to the Illum. Clin. Orthop Rel Res, 162:150-164, 1982.

7. Woolson ST, Harris WH. Complex Total Hip Replacement for Dysplastic or Hypoplastic Hips Using Miniature or Microminiature Components. J. Bone Joint Surg., 65-A:1099-1108, 1983.

8. Harris WH. Autografting and Allografting in Aseptic Failure of Total Hip Replacement. In: The Hip, Proceedings of the 12th Open Scientific Meeting of the Hip Society, Ed. R.B. Welch, St. Louis: CV Mosby Co., 286-295, 1984.

9. Gerber SD, Harris WH. Femoral Head Autografting to Augment Acetabular Deficiency in Patients Requiring Total Hip Replacement. J. Bone Joint Surg., 68-A: 1241-1248, 1986.

10. Harris WH. Bone grafting for acetabular deficiency in association with total replacement. In: The Hip, Proceedings of the Fourteenth Open Scientific Meeting of the Hip Society, CV Mosby Co., Ed. RA Brand, St. Louis: 94-119, 1986.

11. Jasty M, Harris WH. Total Hip Reconstruction Using Frozen Femoral Head Allografts in Patients with Acetabular Bone Loss. Orthop. Clin. of N. Am. 18:291-299, 1987.

12. Mulroy RD Jr., Harris WH. Failure of acetabular autogenous grafts in total hip arthroplasty. Increasing incidence: a follow-up note. J. Bone Joint Surg., 72-A: 1536-1540, 1990.

13. Harris WH. Bulk versus Morselized Bone Graft in Acetabular Revision Total Hip Replacement. Seminars in Arthroplasty 4:68-71, 1993.

14. Kwong LM, Jasty M, Harris WH. High failure rate of bulk femoral head allografts in total hip acetabular reconstructions at 10 years. J. Arthroplasty 8:341-346, 1993.

15. Shinar AA, Harris WH. Bulk structural grafts and allografts for reconstruction of the acetabulum in total hip arthroplasty. Sixteen-year average follow-up. J Bone Joint Surg. Am. 79A:159-168, 1997.

16. Harris WH. Total hip arthroplasty in the management of congenital hip dislocation. Ch. In: The Adult Hip. Ed. Callaghan, JJ, Rosenberg AG, Rubash HE. Lippincott-Raven Publishers, Philadelphia, 1165-1182, 1998.

17. Paavilainen T, Hoikka V, Solonen KA. Cementless Total Replacement for Severely Dysplastic or Dislocated Hips. J Bone Joint Surg/Br 7B, No.2: 205-211, 1990.

18. Jasty M, Anderson MJ, Harris WH. Total Hip Replacement for Developmental Dysplasia of the Hip. Clin. Orthop. Rel Res., 311:40-45, 1995.

19. Woolson ST, Harris WH. A Method of Intraoperative Limb Length Measurement in Total Hip Arthroplasty. Clin. Orthop Rel Res, 194:207-210, 1985.

20. Jasty M, Webster W, Harris WH. Management of limb length inequality during total hip replacement. Clin. Orthop Rel Res, 333:165-171, 1996.

21. Bal BS, Maurer T, Harris WH. Revision of the Acetabular Component without Cement after a Previous Acetabular Reconstruction with Use of a Bulk Femoral Head Graft in Patients who had Congenital Dislocation or Dysplasia: A Follow-up Note. J Bone Joint Surg. 81-A:1703-1706, 1999.

22. Dearborn JT, Harris WH. Acetabular Revision after Failed Total Hip Arthroplasty in Patients with Congenital Hip Dislocation and and Dysplasia: Results After a Mean of 8.6 Years. J Bone Joint Surg. 82-A: (8), 1146-1153, 2000.

23. Russotti GM, Harris WH. Proximal Placement of the Acetabular Component in Total Hip Arthroplasty. A Long-Term Study. J Bone Joint Surg., 73-A:587-592, 1991.

24. Schutzer SF, Harris WH. High placement of porous-coated acetabular components in complex total hip arthroplasty. J. Arthroplasty 9(4):359-367, August 1994.

25. Dearborn JT, Harris WH. High Placement of an Acetabular Component Inserted without Cement in a Revision Total Hip Arthroplasty: Results after a Mean of Ten-Years. J Bone Joint Surg. 81-A (4):469-480, 1999.

26. Bozic KJ, Freiberg AA, Harris WH. The High Hip Center. Clin. Orthop Rel. Res 420: 101-105, 2004.

27. Hendricks KJ, Harris WH. High placement of noncemented acetabular components in revision total hip arthroplasty. A concise follow-up, at a minimum of fifteen years, of a previous report. J. Bone Joint Surg. 88-A(10):2231-2236, 2006.

28. Harris WH. Revision Total Hip Arthroplasty: Reconstruction at a High Hip Center in Acetabular Revision Surgery Using a Cementless Acetabular Component. Orthopedics. 12(9); 991-992, 1998.

29. Harris WH. Management of the Deficient Acetabulum Using Cementless Fixation without Bone Grafting. Orthop. Clin. N. Amer. 24:663-665, 1993.

30. Schmalzried TP, Wessinger SJ, Hill GE, Harris WH. The Harris-Galante Porous Acetabular Component Press-fit without Screw Fixation. J. Arthroplasty 9:235-242, 1994.

31. Anderson MJ, Harris WH. Total Hip Arthroplasty with Insertion of the Acetabular Component without Cement in Hips with Total Congenital Dislocation or Marked Congenital Dysplasia. J Bone Joint Surg. 81-A (3):347-354, 1999.

32. Hampton BJ, Harris WH. Primary Cementless Acetabular Components in Hips with Severe Dysplasia or Total Dislocation. A Concise Follow-Up, at an Average of Sixteen Years, of a Previous Report. J Bone Joint Surg.; 88-A (7):1549-1552, 2006.

33. Muratoglu OK, Imlach H, Jasty M, Harris WH. A new method to determine the locus of radiation damage in retrieved ultra-high molecular weight polyethylene (UHMWPE), In: Characterization and properties of ultra-high molecular weight polyethylene. Gsell, RA, Stein, HL, Ploskonka JJ, Eds. ASTM, 79-94, 1998.

34. Bragdon CR, Jasty M, Muratoglu OK, O'Connor DO, Harris WH. Third-body wear of highly crosslinked polyethylene in a hip simulator. J. Arthroplasty, 18(5):553-561, 2003.

35. Muratoglu OK, Greenbaum ES, Bragdon CR, Jasty M, Freiberg AA, Harris WH. Surface analysis of early retrieved acetabular polyethylene liners: a comparison of conventional and highly crosslinked polyethylenes. J. Arthroplasty, 19(1): 68-77, 2004.

36. Greenbaum ES, Burroughs BB, Harris WH, Muratoglu OK. Effect of lipid absorption on wear and compressive properties of unirradiated and highly crosslinked UHMWPE: an in vitro experimental model. Biomaterials 25(18):4479-4484, 2004.

37. Harris WH. Highly Crosslinked, Electron-Beam-Irradiated, Melted Polyethylene: Some Pros. Clin. Orthop Rel Res, (429): 63-67, 2004.

38. Bragdon CR, Barrett S, Martell JM, Greene ME, Malchau H, Harris WH. Steady-State Penetration Rates of Electron Beam-Irradiated, Highly Crosslinked Polyethylene at an Average 45-Month Follow-Up. J. Arthroplasty; 21(7): 935-943, 2006.

39. Plank GR, Estok DM 2nd, Muratoglu OK, O'Connor DO, Burroughs BR, Harris WH. Contact stress assessment of conventional and highly crosslinked ultra high molecular

weight polyethylene acetabular liners with finite element analysis and pressure sensitive film. Journal of Biomedical Materials Research. Part B, Applied Biomaterials. 80(1):1-10, 2007 Jan.

40. Long WJ, Noiseux NO, Mabry TM, Hanssen AD, Lewallen DG. Uncemented Porous Tantalum Acetabular Components: Early Follow-Up and Failures in 599 Revision Total Hip Arthroplasties, Iowa Orthop J. 35:108-113, 2015.

41. Whitehouse MR, Masri BA, Duncan CP, Garbuz DS. Continued good results with modular trabecular metal augments for acetabular defects in hip arthroplasty at 7 to 11 years. Clin Orthop Rel Res;473(2):521-527, 2015.

42. Makinen TJ, Kuzyk P, Safir OA, Backstein D, Gross AE. Roles of Cages in Revision Arthroplasty of the Acetabulum. J. Bone Joint Surg (AM) Feb 3:98(3):233-242. doi:10.2106/JBJS.O.00143, 2016.

43. Amenabar T, Rahman WA, Hetaimish BM, Kuzyk PR, Safir OA, Gross AE. Promising Mid-term Results with a Cup-Cage Construct for Large Acetabular Defects and Pelvic Discontinuity. Clin Orthop Rel Res. Feb;472(2):408-414, 2016. ls. 80(1):1-10, January 2007.

# CHAPTER 11

---

# PERIPROSTHETIC OSTEOLYSIS: THE DIAGNOSIS AND CURE

IN RETELLING THIS IMPORTANT STORY, it is remarkably difficult to convey the astonishment, surprise and apprehension that swept over me in response to my initial viewing in 1974 of those x rays. The x rays portrayed the left femur of a middle-aged man seven years after his total hip replacement. The destruction of his femur around his left total hip prosthesis was massive.

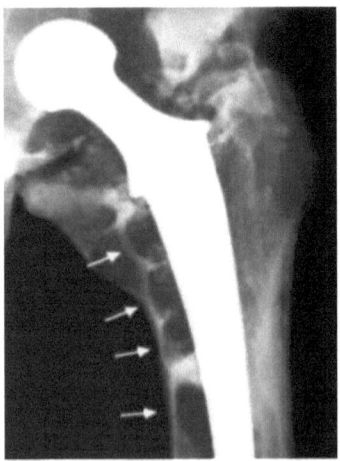

Figure 11.1. The arrows point to multiple regions of massive bone destruction in this AP film of a patient seven years after his total hip replacement, the first case I saw.

Yes, I had previously encountered unsuspected malignancy destroying other areas of the skeleton in other patients, but never immediately around a total hip prosthesis like this. It simply was not part of my long experience in a practice focused solely on patients with arthritis of the hip. Striking was the similarity of this bone destruction in the upper part of his femur around this prosthesis to that of bone destruction caused by a secondary cancer which had spread from some other organ within the body. Against the idea that this might be a primary malignancy was the fact that this site in the skeleton is an uncommon area for primary tumors to arise, even less common in middle-aged men, and virtually unheard of around a prosthesis.

One specific feature of his picture of bone destruction was the failure of his body to react to it by producing new bone. In many circumstances of bone damage or destruction, the body responds by forming new bone, as commonly occurs in cases of infection, injury, and certain types of cancers. Still, it is true that other cancers, either primary or secondary, will destroy bone and yet not prompt any bony reaction. Taken all together, the prime suspect in this unusual case had to be cancer. We would need to garner the wisdom of the entire staff on a problem this unusual by showing the case to Grand Rounds.

Alas, to no avail. No one had ever seen such a picture, particularly since all of our preliminary testing had failed to establish any cancer, or any other disease for that matter, in his body. What in the world was this disease? And what should we do about it?

We needed a biopsy. The overwhelming probability was that direct examination of the tissue would reveal the true nature of the bone destroying process, and statistically the most likely outcome would be simply that of an uncommon example of a common disease, most likely secondary cancer.

But the confusion was compounded by the biopsy, not resolved by the biopsy. The pathologist, likewise, reported something he had never seen before, "sheets of macrophages and many osteoclasts".

What in the world was this combination of macrophages and osteoclasts?

And so it was that we were dumbfounded to see something none of us had ever seen, including the pathologist – a totally unique disease, one never observed before in the long history of mankind.

Brand new diseases always shock, especially late in the 20th century. It simply was not a cancer. The dominant cell type was macrophages – lots of macrophages. When this macrophage-laden tissue lay near bone, the osteoclasts were actively removing bone.

We had learned a lot, but not enough. We now had to discard our prior hypothesis that this was a case of cancer, but what diagnosis should be put in its place? No one on the entire orthopaedic staff had ever seen a disease consisting of sheets after sheets of macrophages and a plethora of very active osteoclasts destroying bone.

The proximity of the total hip implant and adjacent bone cement immediately suggested some sort of causal relationship, but what? Concerns about such massive joint replacement implants causing cancer had long existed, but the evidence supporting this was scant to none, and, of course, this reaction was not malignant anyway.

Unknown at that time was whether or not this disease process itself was painful. Yes, he had pain but also the thigh piece of his total hip had become loose, which in itself is painful. The pain pattern had a mechanical element to it, i.e., it was worse when he walked on his hip, so the pain was likely caused by the loose prosthesis. In any event, because the thigh piece was loose and his hip was sharply painful, he needed a reoperation.

That forced the dilemma. What to do? The proximity of the implant to the bone destruction supported the speculation that some kind of an interrelationship existed between the implant and the bone loss, despite our profound ignorance of what that connection might be. But if somehow, the implant caused the bone destruction, how would redoing the same operation help? And yet, what else could we do? Because the total hip operation mandated discarding the femoral head, none of any of the three alternative hip operations – fusion, osteotomy or cup arthroplasty – would be possible. Simply stated, there was no alternative except to repeat the same total hip operation that he had had before. And there was no option but to use bone cement again. And so, in the face of all these uncertainties, I elected to re-do his total hip with yet another identical total hip replacement. (As an aside, that desperation move served him very well for the rest of his life.)

I had hardly returned to a state of semi-equanimity following this distressing episode when a similar case appeared, and shortly thereafter two others. The biopsies of these cases were all the same, "sheets of macrophages with many osteoclasts". Especially alarmingly, the second and third were cases I had done, making it much more <u>my</u> problem, not just a problem sent to me from afar.

Still, since total hips using bone cement and polyethylene had been done in large numbers around the world since 1962, what I was seeing was a rare disease. This was somewhat reassuring. Or so it seemed.

In fact, this condition was so unusual that the Journal of Bone and Joint Surgery published our manuscript describing just these four cases. To publish just four cases of a problem is very uncommon and clearly was done simply because this problem was so rare, so poorly understood and so alarming.

Uncommon as it was, it turned out that our observations were not unique. Small numbers of critical observers around the world were also seeing rare examples of this bone-destroying process. Since all four of our cases had been successful total hips for a number of years before the trouble had appeared, perhaps <u>duration</u> after the inserting implant might be a factor. Both because of that and because John Charnley had been the first surgeon to use bone cement and also the first to use polyethylene in a total hip, he had both the longest and largest experience in the world. It seemed reasonable to inquire if Charnley had encountered this process.

And indeed he had. But, importantly, under different circumstances. 1962 is often used as the date of Charnley's initiation of the modern era of total hip replacement surgery. But, in fact, his first major effort of this type was in 1957, when he started a large series (nearly 300) of total hip replacements using a different polymer at the articulation, called PTFE, similar to Teflon.

The <u>extraordinary</u> early success of these PTFE cases proved to represent a devastating combination: both exhilarating and short-lived. And in just a few short years nearly all of these cases required removal of the failed total hip, failed because of severe wear of the Teflon-like plastic. Crucially, among some of these cases Charnley found marked bone destruction! And that bone destruction exhibited the identical biopsy pattern that we were seeing sixteen years later, specifically macrophages and osteoclasts.

How had that similarity with our cases become forgotten? Three reasons existed which obscured a direct link between <u>that</u> 1958 experience and what we were seeing. First, Charnley wisely had not released his PTFE prostheses for anyone else to use. So, in fact, Charnley was the only one in the world with that experience.

Secondly, his subsequent new plastic, ultra high molecular weight polyethylene, had replaced the PTFE so successfully that total hips done using polyethylene seemed to be immune to the bone destruction, at least throughout the early years after the insertion.

A third factor lay in his belief, unproven, that such bone destruction was caused by <u>infection</u>. This unusual idea existed in his mind because his infection rate was so high, nearly 10%. So he dismissed further considerations about why the bone was being destroyed because he believed the bone destruction was a consequence of infections –infections, however, that he could not prove, nor even identify. Thus, for these three reasons, our problem in 1974 was not linked to his prior PTFE experience from 1958, at that time,

Alarm over this strange form of failure of total hip replacements multiplied as the incidence grew, compounded by the increasing realization that longer duration of time in service did foster the rising incidence. Dismay was reinforced by both the frustration arising from the complete lack of any explanations and the poor outcomes of the reoperations for this condition. And many of the failures of the reoperations were the direct result of the massive loss of bone, bone which was needed to enable success in the revision surgery. The number of cases with this bone destruction continued to rise and to rise further, until the incidence ultimately peaked at an estimated 1 million patients worldwide. So much for a <u>rare</u> disease.

The terms used to describe the xray appearance of this bone destruction took on strange forms. For example, some called the local erosion of bone in the femur "cavitation". And after Willert advanced his postulate, much, if not all, of this bone destruction was called "cement disease". Also, a dark line appearing on radiographs at the interface between acetabular bone cement and the pelvis was designated as "acetabular demarcation". Demarcation could develop in femurs as well. Because this bone destruction occurred only adjacent to the prosthesis, it was named "periprosthetic osteolysis".

It was not until our study of autopsy-retrieved specimens that the true mechanism of "demarcation" around the acetabular cement was revealed. The acetabular demarcation reported in Charnley's 12-15 year radiographic study had to represent resorption of bone adjacent to the acetabular cement. When this process was progressive, the acetabular component and its entire attached cement mass would become completely loose from the bone. Many acetabular components migrated into a new position as the bone was eaten away. Still, the <u>mechanism</u> of this resorption and thus the "demarcation" mechanism remained a deep mystery.

Our examination of the acetabular region from pelvic specimens which we had retrieved from autopsies unraveled that mystery. Here, too, the bone destruction was caused by the migration of tiny particulate debris, predominantly particulate polyethylene, into the interface between the cement and the bone. This was a startling finding. In fact, at the leading edge of the bone resorption, the microscopic analysis of these autopsy-retrieved acetabular components revealed a "cutting cone" of bone resorption consisting of macrophages ingesting the particulate debris and osteoclasts resorbing the bone. This process, basically similar in nature to the erosive or "balloon" periprosthetic osteolysis that produced "cavitation" in the femur or pelvis, was a manifestation of a linear form of bone reabsorption, and could progress either in a line along the bone-cement interface or could erode expansively, evidencing the common features of the bone resorption elsewhere.

Strangely, however, from the clinical point of view, many of these cemented acetabular components continued to function very well for the patient, despite being loose, even after definite migration. Eventually, however, in increasing numbers, these loose acetabular components would ultimately become painful and thus a clinical failure requiring reoperation.

Demarcation increased in both frequency and extent with increasing time. In the longer studies, by 20 or 30 years after surgery, the percentage of loose acetabular components and revised acetabular components rose to be in the 20-30% range, while among the best cemented femoral components, the number of failed components requiring revision could be ¼ or ⅕ of those figures. Something had to be done.

The total hip research community embarked on a massive medical detective quest to unravel this enigma.

The first break in this painfully slow effort came from Hans Willert, a German orthopaedic professor who had, in addition, long years of training in pathology. He postulated in 1974 that the macrophages had been mobilized to these sites near the prosthesis in order to ingest and potentially eliminate very tiny particles of the bone cement. And indeed, he produced good evidence supporting this theory. The small particles of bone cement arose from beads of cement that became detached from the bulk cement mass or from fragmentation of the bone cement from mechanical failure of the material by fatigue.

With this striking explanation, hip surgeons rejoiced in having an explanation, even though the follow-up question – why and how do macrophages lead to bone destruction – remained unexplained.

Activity at the MGH was the first to initiate efforts to understand the molecular biology of this disease. I recruited Steve Goldring from the Department of Rheumatology and Andrew Rosenberg from Pathology to join me in attempting to unravel this conundrum. We initiated our attack on the problem by showing that the membrane producing the disease in humans elaborated two key compounds, PGE2 and collagenase, as well as having the power to resorb or remove bone when experimentally placed on the skull of a mouse. Subsequently the massive complexity of this most elaborate biologic response to tiny particles was defined further in exquisite detail by many, many other investigators around the world. Moreover, extensive, exquisite molecular biology research from many laboratories showed that it was the activation of the macrophages themselves which ultimately stimulated the osteoclasts.

Willert's postulate created an urgent need to eliminate bone cement. This prompted extensive research in finding a "cementless" way to fix the components to the skeleton. These efforts succeeded by developing total hip components which used a porous layer of metal on the outside of the prosthesis, to allow bone to grow into that layer, locking the implant to the skeleton.

With the increasing demonstrations of success of cementless approaches, a major shift in surgical technique of total hip replacement surgery occurred around the world, with many surgeons espousing this cementless approach with gusto.

This exuberance, however, was acutely shattered when Jorge Galante, John Callaghan, Bill Maloney, and I reported in 1990 our findings of the identical bone destruction occurring in cementless total hip replacements. Lysis could occur without cement! Back to the drawing board! Disillusion compounded the disappointment among patients, hip surgeons and investigators when this disturbing observation was widely confirmed.

But much had been learned from the concept of "cement disease" and it soon followed from the appearance of periprosthetic osteolysis in cementless THR that the second culprit was identified – tiny particles of polyethylene. So, at last the realization that three entirely different particles, namely particles of bone cement, of PTFE and of polyethylene, were all capable of stimulating the bone destruction. In effect, Willert's concept of "cement disease" was not wrong, it was simply incomplete. The true nature of the process was "particle disease", and the tiny particles could consist of any of the different materials used in total hip prostheses, as it was shown later that particles of ceramic and particles of metal could do the same damage.

This was astonishing and disruptive. Clearly the concept of cement disease was both true and incomplete. And the movement to adopt cementless implants and techniques could no longer be expected to eliminate the periprosthetic osteolysis. The correct interpretation of the phenomenon was now "particle disease" and moreover it soon became clear that the particles of the polyethylene from the articulation were far more important than cement particles. What a huge reversal in our understanding of the problem.

Thus, in addition to eliminating bone cement, a critical need existed to eliminate the tiny particles of polyethylene.

What other materials might serve at that articulation without producing as much particulate debris? For decades two "alternate bearings" had been used, but on a limited scale. They were metal-on-metal bearings and ceramic-on-ceramic bearings. Over these decades, while both materials had been improved considerably, increasing their durability and properties, nevertheless both had their own unique disadvantages and had had limited acceptance. Both materials had one special attraction, namely being successful while using larger head diameters, with the specific advantages of larger heads in terms of better range of motion and reduced risk of dislocation.

The third possible pathway was to create a better polyethylene. Here, however, the track record was dismal. All three prior attempts in this regard, all created by major implant manufacturers, had been complete failures and had caused extensive harm. Zimmer had advanced so-called "black poly", a carbon-reinforced polyethylene. Its record was disastrous. Howmedica touted "heat pressed" polyethylene, which had, similarly, harmed many patients. And DePuy extolled "Hylamer", which oxidized prematurely, creating abundant particles which became one more major stimulus to progressive bone destruction.

One remaining, unexplained observation continued to both challenge me and, like all unexplained observations, to offer the possibility of creating opportunity. Ultra high molecular weight polyethylene is widely used in industry for many other purposes and when subjected to wear, produces large flakes, not tiny particles. Why is wear so different in our application? Why was the wear mechanism different when used for total hip articulations compared to industrial uses?

Fortunately, I had previously instituted in the Harris Orthopaedic Lab a retrieval program asking certain carefully selected patients to allow me to retrieve their successful total hip implants after death, for use in our scientific investigations. We had already learned an enormous amount about both the success and the failure of total hip replacement surgery from these returned total hips. And now, we examined this special collection of retrieved specimens to investigate this puzzling problem of the unusual way that wear occurred in ultra high molecular weight polyethylene during use in total hip patients.

Indeed, this avenue proved critical. The findings reflected the perceptive motto proffered by Pasteur on his reflections on a lifetime of scientific investigations, "chance favors only the prepared mind". Our study of 128 retrieved acetabular components, including many from our autopsy retrieval program, provided the key. Using scanning electron microscopy we found a remarkable change in the molecular structure of the polyethylene, a change which had been produced by the continual back and forth motion of gait. As manufactured, polyethylene has no preferred orientation of its extremely long molecules. However, after prolonged human use in total hip arthroplasty, a small region of the molecules in the weight-bearing area became <u>oriented,</u> in the direction of flexion-extension. Based on this unique observation, we postulated our tentative hypothesis. We proposed that this reorientation created the novel wear mechanism that led to the generation of these tiny polyethylene particles. Perhaps, if we could prevent this reorientation, we could reduce the wear. Were that to be true, the follow-on series of questions were:

> Could this reorientation of the molecular structure
> be prevented?

If prevented, would wear be reduced?

If wear were reduced, would it be at any other cost or disadvantage growing out of the method of prevention of the orientation, which would negate or prohibit the advantage?

If wear were safely reduced, would the extent of the reduction be sufficient to bring about a major decrease in the bone destruction?

Would there be major manufacturing or cost impediments?

A still larger question remained. How should we choose to direct our efforts among these three alternative bearings, metal-on-metal, ceramic-on-ceramic and an improved polyethylene? These dilemmas set the stage for our foray into the unknown, the quest to reduce wear at the articular surface of total hip replacement.

The fundamental decision concerning which of the three routes we would choose for generating an improved alternate bearing was inordinately difficult. But in the process of trying to decide "Where to start?", I was diverted totally by the intrusion of a completely different question, "How to start?"

The key question I asked myself was "Even if I invent a superior material, how will I know that it is better?" And perhaps more compelling, "How will I prove its superiority to others?" In short, our first challenge suddenly became to invent a better hip simulator that replicated closely the wear challenge of the human hip, so we could quantitatively assess the efficacy of any new material against wear. Although we launched our efforts to defeat periprosthetic osteolysis in 1990, in fact, the first three years did not involve material research at all. Instead all of our time, effort and intellectual input focused on building this more rigorous hip simulator.

The theoretical demands of a hip simulator were high since ideally the hip simulator should meet both the physical and environmental conditions of human gait. Loads as high as five times body weight had to be cycled in coordination with the gait pattern under normal conditions of joint temperatures and lubrication.

Most of the extant hip simulators left much to be desired, and all were severely limited in stability, requiring a technician in attendance full time. Because of this fragility, they could function only eight hours a day despite such a great need for very prolonged testing. Since many people walk 2 million steps per year and since two steps per second is a rapid walking rate, the requirement that testing should not exceed the normal walking rate sharply limited the number of testing cycles possible in an eight hour day. In fact, the FDA requirements for hip simulator data stipulated only 5 million cycles – despite the fact that for many people this would represent only 2.5 years of walking. That limited requirement was grossly insufficient because periprosthetic osteolysis rarely appeared before five years after insertion and even then, progressively increased in both extent and frequency over the decades thereafter.

To remedy these multiple deficiencies we designed our first hip simulator (See Figure 11.2) containing all these key features.

A load cell on each station

A robust 12-station machine capable of 24-hour service without an attendant

Wide load variability with sophisticated correlation of the load cycle with hip motion

6 degrees of freedom of motion

Upright placement of the total hip components

A pseudocapsule made of flexible plastic to contain the bovine serum used as lubricant

Refined temperature control

**Figure 11.2. Multistation Innovative Hip Simulator, now the world's standard.**

When our new hip simulator was completed, our initial test was against conventional polyethylene. To our utter dismay, the conventional polyethylene did not wear at all. The hip simulator was a complete failure. It did not simulate.

Our disappointment was compounded further by the complete failure of all our subsequent attempts to induce modification of any of our parameters, such as load, frequency, temperature and duration, to produce any wear. Clearly, our design was deficient. In our search for the reason, we turned to the issue of the pattern of gait that we used. Various prior simulators used widely different patterns, ranging from simple flexion and extension to complex circular or elliptical motions which were not related to any normal walking patterns at all.

In our initial iteration, we had followed the advice of a wear expert at MIT who had recommended that simple flexion-extension would be quite satisfactory. However, our failure forced us to reexamine the gait pattern in detail. In so doing I asked the apparently simple question, "What actually happens inside the hip joint during gait? Is the pattern of motion of a given point on the front of the femoral head identical to that on the back, etc. etc.?"

Since this question had never been asked before, we were forced to garner the answer ourselves. This we did in two ways, experimentally by placing a sharp spike at one site on the femoral head and observing the pattern it cut in the polyethylene while going through one full gait cycle. But now, for the first time ever, we included the addition of abduction-adduction with internal-external rotation during the flexion-extension cycle as well as the highly complex correlation of these motions with the proper timing of the variations in load load during the gait cycle. Fortunately, we had built 6 degrees of freedom into the capability of our hip simulator. With this wide capacity we could effect the complex integration of all the load phases with the complex direction of motion involved in gait, i.e., the flexion-extension, abduction-adduction and rotation with the exact load amount appropriate for that instant in gait, into each cycle. Concomitantly, we ran an alternate but parallel investigation of this same question using 3-D computational modeling.

The data from these two disparate types of tests reinforced each other. The pattern of any given point on the femoral head during one gait cycle is roughly a parallelogram but with two totally unanticipated features – the pattern traced by one point overlaps the pattern of an adjacent point and moreover the specific shape, size and direction of any individual parallelogram may vary widely by position on the femoral head. And obviously beyond that, individual aspects of gait versus other diverse activities such as stair climbing, or sit versus stand motions, etc. were quite important. See Figure 11.3.

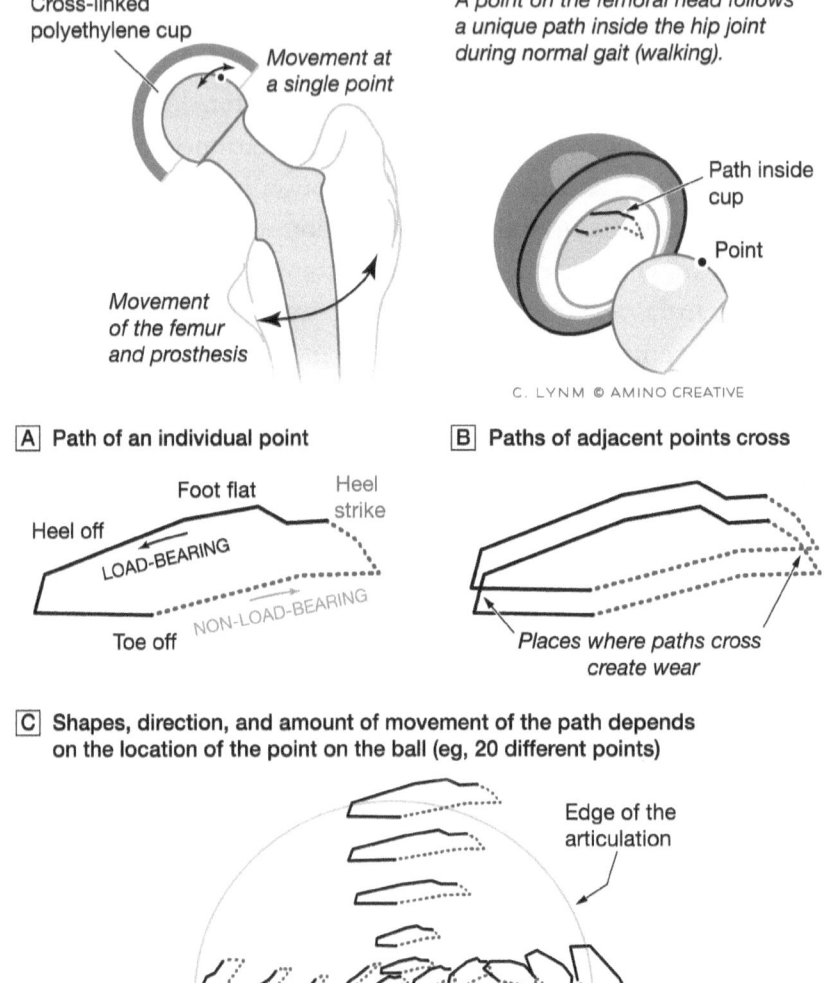

Cross-linked
polyethylene cup

Movement at
a single point

Movement
of the femur
and prosthesis

A point on the femoral head follows
a unique path inside the hip joint
during normal gait (walking).

Path inside
cup

Point

C. LYNM © AMINO CREATIVE

A Path of an individual point

B Paths of adjacent points cross

Foot flat

Heel
strike

Heel off

LOAD-BEARING

NON-LOAD-BEARING

Toe off

Places where paths cross
create wear

C Shapes, direction, and amount of movement of the path depends
on the location of the point on the ball (eg, 20 different points)

Edge of the
articulation

Center of the
articulation

**Figure 11.3. The unexpected pattern of motion of an individual point on the ball of the hip during one gait cycle.** Illustrations © Amino Creative. Printed with permission.

Eureka. Using this new sophisticated gait-load pattern, the wear of the conventional polyethylene matched exactly the wear rate found during human use. Now we needed to develop a wear-resistant material.

This imponderable generated sleepless nights but not much clarity. In the final analysis my decision hinged largely on four features. The first grew out of my own practice, from my own experience, in preferring metal-on-polyethylene joints over metal-on-metal THR. Also, I had concerns about some of the existing shortcomings of both the ceramics and the metal on metal joints. The fourth factor was the fact that metal-on-polyethylene THR were "more forgiving" than the other two types. That particular expression refers to the fact that the metal-on-polyethylene implants are less exacting in such issues as the need for complete accuracy of placement and that they caused less damage to the ball in cases of dislocation of the hip. Specifically, the head can become scratched when a ceramic or metal total hip dislocates compared to none or less head damage if a metal-on-polyethylene total hip dislocates.

Still, it was a hard decision. Based on these "soft" considerations, I chose to commit our efforts to metal-on-polyethylene. And finally, our hypothesis, although theoretical only, had appeal. How would one make polyethylene better, particularly in view of the three seriously flawed attempts by three major orthopaedic manufacturers? Our hypothesis suggested we try to block the reorientation of the strands of the polyethylene.

Then the next set of questions became, how would you achieve that result, that blocking of the reorientation? Would the FDA approve this change? Would it be possible to interest any implant manufacturer in assuming the risks, both economic and legal, in marketing this unknown product worldwide?

The presence of MIT places huge resources close at hand for many issues and particularly issues of material science and especially for polymers. Ed Merrill, Professor of Chemical Engineering at MIT, was already linked to our laboratory, supervising a graduate student working in our lab on bone cement.

My question to him was specific, "Given this unusual reorientation of the molecules of polyethylene by the force of human gait that we have discovered in the polyethylene used in total hips, could you prevent the reorientation?" His answer was characteristic and simple: "Sure." He would do it by crosslinking the polyethylene with electron beam irradiation. Might we be able to successfully negotiate our way through all those follow-on obstacles? Improbable, but exciting.

To accomplish this proposal I hired for the Harris Orthopaedic Lab a very creative, energetic and innovative newly-minted PhD graduate of the MIT Program in Polymer Science and Technology, Orhun Muratoglu. He integrated and supervised our research with that of Ed Merrill and Ed's graduate student at MIT, where the electron beam radiation was done. Orhun's intimate familiarity with MIT from his PhD research made the integration remarkably effective. So too did his motivation, vision and scope.

The introduction of this approach to our specific polymer, ultra high molecular weight polyethylene, did not proceed smoothly. Evidence of trouble among the early trials appeared first in the form of "melting". Melting of this polyethylene differs greatly from the usual concept of melting as it occurs in the transition from ice to water. In "melting" this polymer, melting means that the polyethylene retains its general form and solidity, expands slightly and changes from being an opaque white solid to becoming a translucent solid.

The consternation caused by this melting of the specimen during electron beam irradiation became far more acute with two subsequent adverse events, when some specimens burst into flames and others exploded.

After considerable angst and experimentation, ultimately the cause of these unsettling phenomena was shown to be adiabatic heating. If heat is used to initiate a chemical reaction but the reaction itself also generates heat, the sum of these two processes, if not carefully controlled, can result in excessive heating, enough to produce melting, combustion, or explosion. In our case, when the heat generated by the actual chemical reaction of the crosslinking was added to that introduced by the electron beam irradiation, the combined heat could become excessive, depending on the rate and the specific conditions. After extensive investigation, a series of complex steps were introduced to progressively reduce the rate of the electron beam irradiation to match and offset the heat generated from the crosslinking itself so that, in the end, full crosslinking could be achieved under conditions sufficiently controlled that none of these adverse effects occurred.

The next hurdle we encountered focused on a by-product of the irradiation, namely the "free radicals" which remained in the polyethylene after the crosslinking had been completed. Their presence leads eventually to oxidation of the polyethylene, with concomitant deterioration of its mechanical properties. We had to obliterate this risk.

Fortunately we were able to prove that melting the crosslinked polyethylene under controlled conditions after it had been crosslinked virtually eliminated the free radicals.

Then, to assess other critical features of fully crosslinked, melted polyethylene, we evaluated multiple tests quantifying the mechanical properties of this new product. They all tested out to be satisfactory for use under the forces and conditions of a total hip, with the exception of a slight reduction in fatigue life. We judged that, since the polyethylene as used total hips is primarily stressed in compression, this slight reduction would be unlikely to be adverse or critical.

Thus equipped, we were ready to challenge our hypothesis. The first true test would be a laboratory assessment of the effect of this crosslinking on wear as measured in our advanced hip simulator.

In sharp contrast with the severely disappointing finding of "no wear" when we tested our first iteration of our hip simulator against conventional polyethylene, this time when assessing the wear properties of crosslinked polyethylene on our advanced version of the hip simulator, the very same finding of "no wear" meant success.

We now had our two basic criteria, a new, very low-wear material and proof of its wear resistance against a demanding, sophisticated new hip simulator.

But another stunning result was hidden within our hip simulator results. The exciting "no wear" result applied to the smallest (22 mm) head diameters and also to the largest (32 mm) head diameter that we had tested. That was revolutionary, because ever since the first plastics had been tried in total hip replacement, larger heads had produced more wear. It was this observation that forced Charnley to abandon his preferred initial head size (41.5 mm) and seek out the smallest possible head size (22 mm). While this small head size reduced wear, it decreased the range of motion and increased the risk of dislocation.

Our unique findings implied that we might be able to move the entire field of total hip surgery toward a more normal head size without causing wear and thereby both enhance the range of motion while decreasing the risk of dislocation. A further, extensive set of experiments confirmed this. All this was a remarkable prospect, if the crosslinked polyethylene could be scaled up in production, and safely and economically brought to market, provided it behaved in real life as it did in the laboratory and did not contain some serious, adverse property not yet revealed by our non-human testing.

We built a series of pin-on-disc wear tests to allow study of wear under this newly discovered crossing pattern so that many different issues could be attacked simultaneously without tying up our hip simulator capacity. We also added more simulators and expanded the rigor of our challenge to the crosslinked material by introducing harsher test methods. In real life, wear is often accelerated because abrasive fragments of metal or bone cement work their way into the joint. This is called "third body" wear, and it had never been studied before.

All these events propelled us out of our comfort zone, toward the ultimate end point, the assessment of our new material in actual human use, with all of its unknowns. This meant leaving patient care, orthopaedic surgery and our in-hospital laboratory. None of these new activities play any role in the practice of orthopaedic surgery.

The first step would be to obtain patent protection, actually a step that should have occurred years before. It had not, because of my erroneous belief that one needed to demonstrate proof of concept before applying for a patent. My dismay at finding out how wrong that idea was, was compounded by the knowledge obtained during our patent search that Harry McKellop and his group were simultaneously pursuing a crosslinked polyethylene for THA. The anxiety of that information centered on the question of whether or not their route to this same end would have priority and, in fact, usurp our entire effort, which was now in its eighth year.

Ultimately it did not, but it was only after many years of anxiety that we were reassured about this threat. The ponderous, tortuous, time-and-money consuming steps of obtaining a patent are well illustrated specifically by our experience in which our 1998 patent application for our crosslinked polyethylene was finally granted in 2014. Fortunately, the "patent pending" feature during the prolonged process effectively allowed us to move forward with strong protection of our idea during much of that time.

The question of how many companies to license with our crosslinked polyethylene brought out deep divisions within the individual inventors in the lab. Some felt that the only way to maximize commitment by an industrial partner was to grant a single, exclusive license. Others held strongly that an advance this revolutionary should be as widely available as possible for all patients and thus should be available freely to all manufacturers. The wisdom of granting some degree of exclusivity to a limited number of manufacturers to allow them to offset the huge costs and risks of such an undertaking was further refined from the extensive experience of MIT in granting licenses for new products. Lita Nelsen, their technology transfer leader, stoutly advised having no more than two but also no less than two partners. This choice focused the exclusivity but maintained competition. She also instructed us that, once licensed, the company was in total charge. For example, if they elected not to use the license, for whatever reason, that would be within their rights! We decided to license one large U.S. orthopaedic manufacturer and one European.

While we pondered this issue, suddenly a sharp change in direction occurred. We became informed that if we wished to approach the FDA as inventors, now was the time, since the FDA had just announced that they would be terminating that pathway in 30 days. They had decided to do this because that pathway was virtually never used. This was because the FDA can only grant approval of <u>devices</u> ready to be used, not of ideas or materials. This was an obstacle for us, since we did not manufacture devices. Yet it seemed to us that gaining exposure to their thoughts and criticisms in advance could be valuable, since ultimately we would need their approval for any resultant product. In addition, should they be favorably inclined, our negotiating position with industry would be greatly enhanced.

The mechanics of FDA review involve submission of all germane data in advance, followed by a one-hour oral presentation and question-and-answer session with their experts. However, our one-hour session was totally disrupted by a bomb threat at FDA headquarters. As we watched in disbelief from a field as far from the FDA building as we could run, it proved to be but a bomb scare.

However, fortune shone on us when the senior FDA official in charge of our meeting elected to cancel the remaining appointments of that disrupted afternoon in order to meet with us further, around a picnic table in that field, to pursue extensive discussion stimulated by their interest in our exciting data set. Many additional questions were asked, all of which we were either able to answer then or subsequently after further experimentation.

With this positive interaction with the FDA, but obviously not specific FDA approval, we licensed Sulzer, the large European implant manufacturer, to use our material.

But even with this interest from the FDA, Zimmer, our choice for a US manufacturer, hesitated. They were not interested. After consideration of all other major US manufacturers, we entered into extensive negotiations with Johnson & Johnson. Suddenly, just three days before we were to sign a contract with Johnson & Johnson, Zimmer requested a license. The decision was difficult. Despite being a complete novice in business decision-making, and with great anguish, I reversed our interaction with Johnson & Johnson and chose Zimmer as the second manufacturer to license.

We signed what we considered the two ideal licenses with two large competing companies, one European and one American.

The next hurdle was "scaling up" our bench process into a full manufacturing process. Scaling up a manufacturing process is a hoary problem, proven once again in our case. It worked well for Sulzer, but for Zimmer it succeeded at times and at other times it failed and in a curious form. Some of Zimmer's batches were excellent and others were not useable.

Orhun Muratoglu flew to Seattle to observe Zimmer's ebeam irradiation process first hand and possibly unwind the dilemma. Because Zimmer's ebeam radiation occurred in an unheated warehouse, the starting temperature of the polyethylene blocks varied seasonally with the ambient temperature. This difference in starting temperature altered the efficacy of the temperature-dependent radiation and thus of the adiabatic heating and thus the resultant quality of the product. Empowered by this critical observation, Orhun was rapidly able to resolve this issue.

But not resolved was yet another major unanticipated disruption, the bankruptcy of Sulzer! The bankruptcy resulted from a totally separate manufacturing change completely unrelated to our polyethylene. Sulzer was forced to issue a recall on 17,000 total hip implants, and ultimately driven into bankruptcy. Sulzer was so compromised by this recall and the class action suit against them that when they came out of bankruptcy, they were vulnerable. Zimmer bought the company. Thus disappeared our preferred strategy of licensing two competing companies.

Our key challenge remained, to prove that the new crosslinked polyethylene was truly a very low-wear material when applied in the ultimate test, human use. But in people, of course, we could no longer remove the polyethylene to measure wear as we did from the hip simulator. Faced with this critical need, we turned to the world-renowned, rigorous group of investigators in Goteborg, Sweden to apply their complex but remarkably accurate specialized xray technique called radiostereogramography or RSA. As improbable as it would seem, RSA can be applied to assess polyethylene wear in humans in real life with a resolution of only 30 microns. In 1998 we arranged for their RSA measurement of wear of our crosslinked polyethylene in their patients in a very unusual, long-term study designed to last over the next 10 years.

It was then that the most dangerous, unspoken risk lurking behind our worst fears surfaced. That fear was the completely unknowable risk of some totally unanticipated, severe, adverse reaction to our crosslinked polyethylene. This apprehension was catapulted forward in our thinking by front page headlines reporting exactly such a feared issue, a massive disaster in one of the other two "alternate bearings". This disaster was the rapid, severely adverse development called "pseudotumor formation" among metal-on-metal total hip replacements. Pseudotumor formation is a characterized by a localized, aggressive, destructive reaction by the body to the particulate debris from tiny metal particles caused by wear. It was even more severe than had occurred with the "particle disease" from total hip replacements using conventional polyethylene.

In many series of metal-on-metal implants, 30% or more of the patients required reoperations after just six years, reoperations which were extremely difficult, with very limited results. Not until 650,000 metal-on-metal total hip replacements had been done in the USA alone was this form of "alternate bearing" total hip replacement surgery abandoned!

Against this alarming backdrop, we were appalled to examine for the first time a crosslinked polyethylene specimen which had been retrieved after it had been in use in a patient. The surface of the polyethylene was badly deformed. It was severely scuffed and distorted in a way we had never seen before. And so were our subsequent retrieved human specimens.

Happily, Orhun resolved our fear, by applying his polymer expertise. Ultra high molecular weight polyethylene exhibits a curious property called "shape memory". If a material has "shape memory", it can be deformed or bent but then if it is melted and subsequently cooled, "shape memory" returns it to its original shape. After melting and cooling, these frightening looking specimens which we had retrieved after use in humans returned fully to their pristine configuration, identical to the day they were made. There was no wear! There had been deformation and scuffing, but no wear. No material had been lost.

Now what are the results of the prolonged, extensive human experience with total hip replacements done using our crosslinked polyethylene? In short, assessment by many different investigators who were studying widely varied populations from multiple different countries around the world reflected the aggregate experience of an estimated 7,000,000 patients covering an interval of 18 years. The results were remarkable. Wear was very low, virtually non-existent. Even more important, bone destruction around the total hip implants has apparently been eliminated. The outlook for both initial success and long durability of total hip replacements has been markedly improved. The risk of failure of a total hip replacement has been reduced by an order of magnitude by crosslinked polyethylene.

For example, in a nationwide study of total hip replacements done in Australia, even after 15 years of use in patients, the total number of reoperations for bone destruction and/or loosening of a component was 1.1%. This means that virtually 99% of the patients, even after 15 years of use, have been free of the risk of reoperation from either a loose component or the bone destruction itself.

This represents a most remarkable success. It is a success of medical discovery concerning the nature of an iatrogenic novel disease, a success of materials innovation and a success of major magnitude in patient care. Some observers have designated it the single most important progress in total hip surgery since its introduction 60 years ago.

## Subsequent Advances

Still, there was much more to be done. Orhun assumed the sole responsibility of developing further materials which might continue such advances. Over time these materials took three forms.

The first arose from a fascinating side issue, not a project specifically relevant to our core activities. As a member of Cambridge Polymer Group (CPG), Orhun began investigation of this problem in conjunction with this polymer start-up company begun by Stephen Spiegelberg, who had been a fellow PhD student in polymer sciences at MIT. CPG had been asked by Boston Scientific to study a compelling and alarming failure, that of the spontaneous fracture of cardiac catheters used in the insertion of coronary artery stents. A similarity to our existing work lay in that these coronary stent catheters were made of ultra high molecular weight polyethylene which had been sterilized by gamma ray radiation. Prior to sterilization, each long catheter had been coiled in an elliptical shape and packaged. Once irradiated in this package, they remained sterile until opened for use.

Interestingly, the fracturing of the ultra high molecular weight polyethylene catheter during use occurred only after prolonged shelf life. This, obviously, raised the question of time-related deterioration of the irradiated polyethylene from oxidation, possibly caused by the free radicals generated as a by-product of the gamma radiation sterilization.

Examination of such catheters after a long shelf life produced a very curious observation. The straight segments of the catheter within its elliptical shape had oxidized heavily while the curved portions of the elliptical shape had not. Further investigation suggested that the deformation of the polyethylene necessary to create the curved segments of the coil, in contrast to the reduced strain occurring in the unstressed straight segments, decreased the concentration of free radicals. After this hypothesis was confirmed by direct experimentation, studies proved that sufficient compression of the polyethylene would eliminate the free radicals.

After extensive development, Orhun was able to use commercially available compression techniques to eliminate those free radicals caused by irradiation. This, in turn, made it possible to delete the need for the melting step from our existing protocol which we had been using to produce our initial crosslinked polyethylene. Therefore, both from the point of view of avoiding the melting step which weakened the polyethylene somewhat and from the point of view of having a distinctly different product, this form of highly crosslinked polyethylene contained specific advantages.

More important was the second major thrust beyond first generation crosslinked polyethylene. Adding vitamin E to the polyethylene was a totally different method of counteracting the residual free radicals created by the irradiation. Thus further valuable advances developed.

Vitamin E, the widely used antioxidant, could be added to either the polyethylene powder or diffused into the solid polyethylene block under elevated temperatures, in order to counteract the free radicals. Through extensive, imaginative investigations Orhun developed several optimizing techniques for creating crosslinked polyethylene containing vitamin E in adequate concentrations to obliterate the free radicals. That this was true in a variety of different forms had multiple advantages. As noted above the vitamin E negated the need to melt the polyethylene to eliminate the free radicals. By so doing, that reduction in mechanical properties which resulted from the melting was eliminated.

Secondly, the persistence of the vitamin E in the material allowed these preparations to offset two curious and newly identified forms of the in vivo generation of new free radicals, namely, those created by lipid migration into areas of the crosslinked polyethylene and those created in response to stress. Thus, the presence of Vitamin E offset these two newly discovered, in vivo sources of late oxidation of our first generation, melted crosslinked polyethylene which arose, in vivo, despite the melting process which had been applied during manufacturing.

Additionally, by creatively applying selected diffusion methods, the location of the vitamin E within the implant could be specified. For example, the vitamin E could be preferentially restricted to a certain depth from the surface of the final product. Thus both the surface and the interior could be specifically and differentially created to optimize different functions. For example, a "sandwich" component was developed, with virgin ultra high molecular weight polyethylene throughout the depth of the interior for better mechanical properties and, yet this core area of virgin polyethylene is contained within a "sandwich" configuration by making the front and back surfaces crosslinked by irradiating only to a certain depth to generate the crosslinks required to decrease wear. Simultaneously the incorporated vitamin E would offset both the oxidation which would ordinarily result from the radiation and that from the later reactions in vivo.

Because of the distinct advantages of both <u>complete</u> and <u>continuous</u> effects against oxidation afforded by the vitamin E and the improved mechanical properties afforded by the avoidance of melting, those second generation crosslinked polyethylenes containing vitamin E exhibited unique appeal among metal-on-polyethylene total joint articulations.

The third major innovation thrust us into the unique area of using crosslinked polyethylene to be a carrier for the delivery of other molecules to the wound area.

Of great potential are the two separate and important possibilities of addressing the severe problems of deep wound infection and of pain relief, based on the general principle of using polymers as the delivery vehicle for prophylactic and/or therapeutic antibiotics and similarly for delivering pain medicine. While a somewhat parallel concept exists for total hip surgery in the original form of Buchholz's creative addition of antibiotics to bone cement, a concept further embellished by Duncan's two-stage reconstruction technique of infected total hips and knees using the PROSTALAC approach, Orhun's new concept expands this delivery approach greatly by using, instead, polyethylene as the drug delivery vehicle.

Both the magnitude of release and the duration of release of antibiotics from the polyethylene can be substantially increased using polyethylene as the reservoir compared to release from bone cement. Thus, for both prophylaxis and for treatment of prosthetic joint infection, these approaches are very attractive. Particularly in dealing with the prevention and the obliteration of resistant infections such as very resistant infections called MRSA and comparable analogs, the potential for dramatic resolution of some of the most challenging remaining problems in THA and TKA is remarkably enticing.

## SELECTED RELATED REFERENCES

McKee GK, Watson-Farrar J. Replacement of arthritic hips by the Mckee-Farrar prosthesis. J Bone Joint Surg. 48B(2): 245-258, 1966.

Boutin P. Total arthroplasty of the hip with aluminum prostheses (article in French). Acta Orthop Belg. 40(5-6): 744-754, 1974.

Harris WH, Schiller AL, Scholler J-M, Freiberg RA, Scott R. Extensive localized bone reabsorption in the femur following total hip replacement. J Bone Joint Surg. 58-A(5): 612-618, 1976.

DeLee J, Charnley J. Radiological demarcation of cemented sockets in total hip replacement. Clin Orthop Rel Res. 121: 20-32,1976.

Willert HG. Reactions of the articular capsule to wear products of artificial joint prostheses. J Biomed Mater Res. 11(2): 157-164, 1977.

Charnley J. Low Friction Arthroplasty of the Hip: Theory and Practice. New York, NY: Springer-Verlag Berlin Heidelberg; 1979.

Wright TM, Fukubashi T, Burstein AH. The effect of carbon fiber reinforcement on contact area, contact pressure, and time-dependent deformation in polyethylene tibial components. J Biomed Mater Res. (15-5): 719-730, 1981.

Goldring SR, Schiller AL, Roelke M, Rourke CM, O'Neill DA, Harris WH. The synovial-like membrane at the bone-cement interface in loose total hip replacements and its proposed role in bone lysis. J Bone Joint Surg.; 65-A: 575-583, 1983.

Jones LC, Hungerford DS. Cement disease. Clin Orthop Rel Res. December 225: 192-206, 1987.

Willert HG, Bertram H, Buchhorn GH. Osteolysis in alloarthroplasty of the hip. Clin Orthop Rel Res. 258: 108-120, 1990.

Maloney WJ, Jasty M, Harris WH, Galante JO, Callaghan JJ. Endosteal erosion in association with stable uncemented femoral components. J Bone Joint Surg. 72-A(7): 1025-1034, 1990.

Schmalzried TP, Kwong LM, Jasty M et al. The mechanism

of loosening of cemented acetabular components in total hip arthroplasty. Clin Orthop Rel Res.; 274:60-78, 1992.

Kwong LM, Jasty M, Mulroy RD, Maloney WJ, Bragdon CR, Harris WH. The histology of the radiolucent line. J Bone Joint Surg. 74(B): 67-73, 1992.

Schmalzried TP, Jasty M, Harris WH. Periprosthetic bone loss in total hip arthroplasty. Polyethylene wear debris and the concept of the effective joint space. J Bone Joint Surg. A(74): 849-863, 1992.

Jiranek WA, Michado M, Jasty M, et al. Production of cytokines around loosened cemented acetabular components. J Bone Joint Surg. 75-A(6): 863-879, 1993.

Kabo JM, Gebhard JS, Loren G, Amstutz HC. In vivo wear of polyethylene acetabular components. J Bone Joint Surg. (Br) 75-B: 254-258, 1993.

Jasty M, Bragdon CR, Jiranek W, Chandler H, Maloney W, Harris WH. Etiology of osteolysis around porous-coated cementless total hip arthroplasties. Clin Orthop Rel Res. 308: 111-126, 1994.

Li S, Burstein AH. Ultra-high molecular weight polyethylene. J Bone Joint Surg. 76-A, 1080-1090, 1994.

Harris WH. The problem is osteolysis. Clin Orthop Rel Res. Feb. 311:46-531995.

Zicat B, Engh CA, Gokcen E. Patterns of osteolysis around total hip components inserted with and without cement. J Bone Joint Surg. 777-A(3): 432-439, 1995.

Smith E, Harris WH. Increasing prevalence of femoral lysis in cementless total hip arthroplasty. J. Arthroplasty. 1995 10: 407-412, 1995.

Harris WH. The problem is osteolysis. Clin Orthop Rel Res. 311:46-53, 1995.

Willert HG, Semlitsch M. Tissue reactions to plastic and metallic wear products of joint endoprostheses. Clin Orthop Rel Res. 333: 4-14, 1996.

Bragdon CR, O'Connor DO, Lowenstein JD, Jasty M, Syniuta WD. The importance of multidirectional motion on the wear of polyethylene. Proc Inst Mech Eng (H). 210(3): 157-165, 1996.

Ramamurti BS, Bragdon CR, O'Connor DO et al. Loci of movement of selected points on the femoral head during normal gait: three-dimensional computer simulation. J Arthroplasty 11: 845-852, 1996.

Schmalzreid TP, Peters PC, Maurer BT, Bragdon CR, Harris WH. Long-duration metal-on-metal total hip arthroplasties with low wear of the articulating surfaces. J Arthroplasty 11: 322-331, 1996.

Chmell MJ, Poss R, Thomas WH, Sledge CB. Early failure of hylamer acetabular inserts due to eccentric wear. J Arthroplasty 11(3): 351-353, 1996.

Premnath V, Harris WH, Jasty M, Merrill EW. Gamma sterilization of UHMWPE articular implants: an analysis of the oxidation problem. Biomaterials 17:1741-1753, 1996.

Livingston BJ, Chmell MJ, Spector M, Poss R. Complications of total hip arthroplasty associated with the use of an acetabular component with a hylamer liner. J Bone Joint Surg. 79(A): 1529-1538, 1997.

Jasty M, Goetz DD, Bragdon CR et al. Wear of polyethylene acetabular components in total hip arthroplasty. An analysis of 128 components retrieved at autopsy or revision operations. J Bone Joint Surg. 79-A: 349-358, 1997.

Muratoglu OK, Imlach H, Jasty M, Harris WH. A new method to determine the locus of radiation damage in retrieved ultra-high molecular weight polyethylene (UHMWPE), In: Characterization and properties of ultra-high molecular weight polyethylene. Gsell RA, Stein HL, Ploskonka JJ, Eds. ASTM, 79-94, 1998.

Hellman EJ, Capello WN, Feinberg JR. Omnifit cementless total hip arthroplasty. Clin Orthop Rel Res. 364: 164-174, 1999.

McKellop H, Shen F, Lu B, Campbell P, Salovey R. Development of an extremely wear-resistant ultra high molecular weight polyethylene for total hip replacements. J Orthop Rel Res. 17: 157-167, 1999.

Duffy GP, Berry DJ, Rowland C, Cabanela ME. Primary

uncemented total hip arthroplasty in patients <40 years old. J Arthroplasty 16(8) Suppl 1: 140-144, 2001.

Muratoglu OK, Bragdon CR, O'Connor DO, Jasty M, Harris WH. A Novel Method of Crosslinking Ultra-high Molecular Weight Polyethylene to Improve Wear, Reduce Oxidation, and Retain Mechanical Properties. Recipient of the HAP Paul Award. J. Arthroplasty 16(2):149-60, 2001.

Harris WH. Wear and periprosthetic osteolysis: the problem. Clin Orthop Rel Res. 393: 66-70, 2001.

Jacobs JJ, Roebuck KA, Archibeck M, Hallab NJ, Glant TT. Osteolysis: basic science. Clin Orthop Rel Res. Dec (393): 71-77, 2001.

Archibeck MJ, Jacobs JJ, Roebuck KA, Glant TT. The basic science of periprosthetic osteolysis. Instr Course Lect. 50: 185-195, 2001.

Muratoglu OK, Bragdon CR, O'Connor D, et al. Larger diameter femoral heads used in conjunction with a highly crosslinked ultra-high molecular weight polyethylene: a new concept. J Arthroplasty 16(8) Suppl 1: 24-30, 2001.

Muratoglu OK, Bragdon CR, O'Connor DO, Jasty M, Harris WH. A novel method of cross-linking ultra-high-molecular-weight polyethylene to improve wear, reduce oxidation, and retain mechanical properties. J Arthroplasty 16(2): 149-160, 2001.

Muratoglu, OK, Harris, WH. Identification and Quantification of Irradiation in UHMWPE through Trans-Vinylene Yield. J Biomed Mat Res.56 (4):584-592, 2001.

Harris WH. Cross-linked Polyethylene: Why the Enthusiasm? Instr. Course Lect. 50: 181-184, 2001.

Crowther JD, Lachiewicz PF. Survival and polyethylene wear of porous-coated acetabular components in patients less than fifty years old. J Bone Joint Surg. 84(5): 729-735, 2002.

Dumbleton JH, Manley MT, Edidin AA. A literature review of the association between wear rate and osteolysis in total hip arthroplasty. J Arthroplasty 17: 649-661, 2002.

Muratoglu OK, Harris WH. Use of highly crosslinked ultra-high molecular weight polyethylene in total hip replacement to decrease the generation of wear debris and reduce periprosthetic osteolysis. Seminars in Arthroplasty. 13(4), 318-324, 2002.

Bragdon CR, Jasty M, Muratoglu OK, O'Connor DO, Harris WH. Third-body wear of highly crosslinked polyethylene in a hip simulator. J Arthroplasty. 18(5): 553-561, 2003.

Muratoglu OK, Ruberti J, Melotti S, Spiegelberg SH, Greenbaum ES, Harris WH. Optical analysis of surface changes on early retrievals of highly cross-linked and conventional polyethylene tibial inserts. J Arthroplasty 18(7) Suppl 1: 42-47, 2003.

Muratoglu OK, Merrill EW, Bragdon CR, O'Connor D, Hoeffel D, Burroughs B, Jasty M, Harris WH. Effect of radiation, heat, and aging on in vitro resistance of polyethylene. Clin. Orthop. Rel Res. (471): 253-262, 2003.

Muratoglu OK, Greenbaum ES, Bragdon CR, Jasty M, Freiberg AA, Harris WH. Surface analysis of early retrieved acetabular polyethylene liners: a comparison of conventional and highly cross-linked polyethylenes. J Arthroplasty 19(1): 68-77, 2004.

Duffy P, Sher JL, Partington PF. Premature wear and osteolysis in an HA-coated, uncemented total hip arthroplasty. J Bone Joint Surg. 86-B(1): 34-38, 2004.

Sanchez-Sotelo J, Lewallen DG, Harmsen WS, Harrington J, Cabanela ME. Comparison of wear and osteolysis in hip replacement using two different coatings of the femoral stem. Int Orthop. 28: 206-210, 2004.

Greenbaum ES, Burroughs BB, Harris WH, Muratoglu OK. Effect of lipid absorption on wear and compressive properties of unirradiated and highly crosslinked UHMWPE: an in vitro experimental model. Biomaterials. 25(18):4479-4484, 2004.

Oral E, Wannomae KK, Hawkins N, Harris WH, Muratoglu OK. Alpha-Tocopherol-Doped Irradiated UHMWPE for High Fatigue Resistance and Low Wear. Biomaterials 25(24):5515-5522, 2004.

Oral E, Wannomae KK, Hawkins N, Harris WH, Muratoglu

OK. Alpha-Tocepherol-doped irradiated UHMWPE for high fatigue resistance and low wear. Biomaterials. 25(24):5515-5522, 2004.

Willert HG, Buchhorn GH, Fayyazi A, Flury R, Windler M, Lohmann CH. Metal-on-metal bearings and hypersensitivity in patients with artificial hip joints. A clinical and histomorphological study. J Bone Joint Surg. 87(1):28-36, 2005.

Bragdon CR, Jasty M, Muratoglu OK, Harris WH. Third-body wear testing of a highly crosslinked acetabular liner: the effect of large femoral head size in the presence of particulate poly(methyl-methacrylate) debris. J Arthroplasty 20(3): 379-85, 2005.

Jasty M, Rubash H, Muratoglu OK. Highly cross-linked polyethylene: the debate is over – in the affirmative. J Arthroplasty 20(4) Suppl 2: 55-58, 2005.

Oral E, Greenbaum ES, Malhi AS, Harris WH, Muratoglu OK. Characterization of Irradiated Blends of Alpha-tocopherol and UHMWPE. Biomaterials 26(33):6657-6663, 2005.

Wannomae KK, Christensen SD, Freiberg AA, Bhattacharyya S, Harris WH, Muratoglu OK. The effect of real-time aging on the oxidation and wear of highly cross-linked UHMWPE acetabular liners. Biomaterials Mar 27(9):1980-7, 2006.

Wannomae KK, Bhattacharyya S, Freiberg A, Estok D, Harris WH, Muratoglu O. In Vivo Oxidation of Retrieved Crosslinked Ultra High Molecular Weight Polyethylene Acetabular Components with Residual Free Radicals. J. Arthroplasty. Oct 21(7); 1005-1011, 2006.

Burroughs BR, Muratoglu OK, Bragdon CR, Wannomae KK, Christensen S, Lozynsky AJ, Harris WH. In vitro comparison of frictional torque and torsional resistance of aged conventional gamma-in-nitrogen sterilized polyethylene versus aged highly crosslinked polyethylene articulating against head sizes larger than 32 mm. Acta Orthopaedica; Oct 77 (5); 710-718, 2006.

Oral E, Wannomae K, Rowell SL, Muratoglu OK. Migration Stability of α-Tocopherol in Irradiated UHMWPE. Biomaterials 27(11): 2434-2439, 2006.

Oral E, Christensen S, Wannomae KK, Malhi A, Muratoglu

OK. Wear Resistance and Mechanical Properties of Highly Cross-linked UHMWPE Doped with Vitamin E. J Arthroplasty 21(4): 580-591, 2006.

Oral E, Rowell SL, Muratoglu OK. The Effect of α-Tocopherol on the Oxidation and Free Radical Decay in Irradiated UHMWPE. Biomaterials 27: 5580-5587, 2006.

Hallan G, Lie SA, Havelin LI. High wear rates and extensive osteolysis in 3 types of uncemented total hip arthroplasty. Acta Orthopaedica. 77(4): 575-584, 2006.

Estok DM II, Burroughs BR, Muratoglu OK, Harris WH. Comparison of hip simulator wear of 2 different highly crosslinked ultra high molecular weight polyethylene acetabular components using both 32- and 38-mm femoral heads. J Arthroplasty 22(4): 581-589, 2007.

Plank GR, Estok DM II, Muratoglu OK, O'Connor DO, Burroughs BR, Harris WH. Contact stress assessment of conventional and highly crosslinked ultra high molecular weight polyethylene acetabular liners with finite element analysis and pressure sensitive film. Journal of Biomedical Materials Research Part B, Applied Biomaterials. January 80(1):1-10, 2007.

Bragdon CR, Greene M, Freiberg AA, Harris WH, Malchau H. Radiostereometric analysis comparison of wear of highly cross-linked polyethylene against 36- vs 28-mm femoral heads. J Arthroplasty September 22(6) Suppl 2: 125-129, 2007.

Oral E, Wannomae KK, Rowell SL, Muratoglu OK. Diffusion of Vitamin E in UHMWPE. Biomaterials 28: 5225-5237, 2007.

Laurent MP, Johnson TS, Crowninshield RD, Blanchard CR, Bhambri SK, Yao JQ. Characterization of a highly crosslinked ultrahigh molecular weight polyethylene in clinical use in total hip arthroplasty. J Arthroplasty 23(5): 751-761, 2008.

Bodugoz-Senturk H, Choi J, Oral E, Kung JH, Macias CE, Braithwaite G, Muratoglu OK. The Effect of Polyethylene Glycol on the Stability of Pores in Polyvinyl Alcohol Hydrogels During Annealing. Biomaterials 29(2):141-149, 2008.

Oral E, Godleski-Beckos CM, Lozynsky AJ, Malhi AS,

Muratoglu OK. Wear Reduction and Fatigue Strength Increase in Vitamin E-containing High Pressure Crystallized Ultrahigh Molecular Weight Polyethylene. Biomaterials 30:1870-1880, 2009.

Bodugoz-Senturk H, Macias CE, Kung JH, Muratoglu OK. Poly (Vinyl Alcohol)-Acrylamide Hydrogels as Load-bearing Cartilage Substitute. Biomaterials 30: 589-596, 2009.

Anseth S, Pulido PA, Adelson WS, Patil SP, Sandwell JC, Colwell CW Jr. Fifteen-year to twenty-year results of cementless Harris-Galante porous femoral and Harris-Galante porous I and II acetabular components. J Arthroplasty 45(5): 687-691, 2010.

Langton DJ, Jameson SS, Joyce TJ, Hallab HJ, Natu S, Nargol AV. Early failure of metal-on-metal bearings in hip resurfacing and large diameter total hip replacement: A consequence of excess wear. J Bone Joint Surg. 92: 38-46, 2010.

Jarrett BT, Cofske J, Rosenberg AE, Oral E, Muratoglu OK, Malchau H. In Vivo Biological Response to Vitamin E and Vitamin E-doped Polyethylene. Journal of Bone and Joint Surgery 92 (16): 2672-2681, 2010.

Muratoglu OK, Wannomae KK, Rowell SL, Micheli BR, Malchau H. Ex vivo stability loss of irradiated and melted ultra-high molecular weight polyethylene. J Bone Joint Surg Am 1;92(17):2809-2816,. Doi: 10.21/JBJS.1.01017, Dec. 2010.

Oral E, Ghali BW, Rowell SL, Micheli BR, Lozynsky AJ, Muratoglu OK. A Surface Cross-linked UHMWPE Stabilized by Vitamin E with Low Wear and High Fatigue Strength. Journal of Biomaterials 31:7051-7060, 2010.

Wannomae KK, Christensen S, Micheli B, Rowell S, Schroeder D, Muratoglu OK. Delamination and Adhesive Wear Behavior of α-Tocopherol–Stabilized Irradiated Ultrahigh-Molecular-Weight Polyethylene. Journal of Arthroplasty 25(4):635-643, 2010.

Rowell SL, Oral E, Muratoglu OK. Comparative Stability of Vitamin E-blended and Diffused Cross-linked UHMWPEs After 3-year Real Time Aging. Journal of Orthopaedic Research 29(5): 773-780, 2011.

Senturk HB, Oral E, Choi J, Macias C, Muratoglu OK.

Molecular Weight Effect of Theta-Gel Formation in Poly(Vinyl Alcohol)-Poly(Ethylene Glycol) Mixtures. Journal of Applied Polymer Science, 2011.

Meier B. Metal hips failing fast, report says. *The New York Times*, Health Section. September 15, 2011.

Malchau H, Bragdon CR, Muratoglu OK. The stepwise introduction of innovation into orthopaedic surgery: the next level of dilemmas. J Arthroplasty Sept. 26(a): 825-831, 2011.

Meier B. Failure of artificial hip is expected to cost billions. *The Boston Globe*, Business Section. December 28, 2011.

Cohen D. Out of Joint: the Story of the ASR. *MJ.* May 14, 2011. doi:http://dx.doi.org/10.1136/bmj.d2905

Kurtz SM, Gawel HA, Patel JD. History and systematic review of wear and osteolysis outcomes for first-generation highly crosslinked polyethylene. Clin Orthop Rel Res. 469: 2262-2277, 2011.

Oral E, Ghali BW, Muratoglu OK. The Elimination of Free Radicals in Irradiated UHMWPEs with and without Vitamin E Stabilization by Annealing under Pressure. Journal of Biomedical Materials Research 97B(1):167-174, 2011.

Micheli BR, Wannomae KK, Lozynsky AJ, Christensen SD, Muratoglu OK. Knee Simulator Wear of Vitamin E Stabilized Irradiated UHMWPE. Journal of Arthroplasty 27 (1): 95-104, 2011.

Bichara D, Zhao X, Hwang NS, Bodugoz-Senturk H, Ballyns FP, Oral E, Randolph MA, Bonassar LJ, Gill TJ, Muratoglu OK. Porous Poly (vinyl alcohol)-Hydrogel Matrix-Engineered Biosynthetic Cartilage. Tissue Eng Part A 17 (3-4): 301-309, 2011.

Oral E, Beckos C, Muratoglu OK. Homogenization of Supercritical Carbon Dioxide Enhances the Diffusion of Vitamin E in UHMWPE. Journal of Applied Polymer Science 124(1): 518-524, 2011.

Mall NA, Nunley RM, Zhu JJ, Maloney WJ, Barrack RL, Clohisy JC. The incidence of acetabular osteolysis in young patients with conventional versus highly crosslinked polyethylene. Clin Orthop Rel Res. 469(2): 372-381, 2011.

Geraint ER, Thomas GE, Simpson DJ, et al. The seven-year

wear of highly cross-linked polyethylene in total hip arthroplasty: a double-blind, randomized controlled trial using radiostereometric analysis. J Bone Joint Surg. 93: 716-722, 2011.

Milner GR, Boldsen JL. Humeral and femoral head diameters in recent white American skeletons. J. Forensic Sci. January; 57(1):35-40, 2012.

Oral E, Neils A, Rowell S, Lozynsky A, Muratoglu OK. Increasing Irradiation Temperature Maximizes Vitamin E Grafting and Wear Resistance of UHMWPE. Journal of Biomedical Materials Research B 101(3):436-440, 2012.

Choi J, Hsiang KJ, Macias CE, Muratoglu OK. Highly Lubricious Poly(Vinyl Alcohol)-Poly(Acrylic Acid) Hydrogels. Journal of Biomedical Materials Research 100B(2): 524- 532, 2012.

Johanson P, Digas G, Herberts P. Highly crosslinked polyethylene does not reduce aseptic loosening in cemented THA. 10-year findings of a randomized study. Clin Orthop Rel Res. 470(11): 3083-3093, 2012.

Engh CA Jr, Hopper RH Jr, Huynh C, Ho H, Sritulanondha S, Engh CA Sr. A prospective, randomized study of cross-linked and non-cross-linked polyethylene for total hip arthroplasty at 10-year follow-up. J Arthroplasty 27(8): 2-8, 2012.

Yeung E, Bott PT, Chana R et al. Mid-term results of third-generation alumina-on-alumina ceramic bearings in cementless total hip arthroplasty. A ten-year minimum follow-up. J Bone Joint Surg. 94(A)2: 138-144, 2012.

Howie DW, Holubowycz OT, Middleton R, and the Large Articulation Study Group. Large femoral heads decrease the incidence of dislocation after total hip arthroplasty. J Bone Joint Surg. 94(A)12: 1095-1102, 2012.

Garbuz DS, Masri BA, Duncan CP et al. Do large heads (36 and 40 mm) result in reduced dislocation rates in a randomized clinical trial? Clin Orthop Rel Res. 470: 351-356, 2012.

Oral E, Neils AL, Lyons C, Fung M, Doshi B, Muratoglu OK. Surface cross-linked UHMWPE can enable the use of larger femoral heads in total joints. J Orthop Rel Res. (1)59-66 Jan 2013: Doi 10.1002/jor.22195. Epub 2012 July 30.

Garcia-Rey R, Garcia-Cimbrelo E, Cruz-Pardos A. New polyethylenes in total hip Replacement: a 10-to-12-year follow up study. Bone Joint Journal. 95: 326-332, 2013.

Goodman SB, Gibon E, Yao Z. The basic science of periprosthetic osteolysis. PMC Instr Course Lect. 62: 201-206, 2013.

Meier B. FDA seeks to tighten regulations of all-metal hip implants. *The New York Times,* January 16, 2013.

Babovic N, Trousdale, RT. Total hip arthroplasty using highly cross-linked polyethylene in patients younger than 50 years with minimum 10-year follow-up. J Arthroplasty *28: 815-817,* 2013.

Bragdon C, Doerner M, Rubash H et al. Clinical multi-centric studies of the wear performance of highly crosslinked remelted polyethylene in THR. Clin Orthop Rel Res. 471(2): 393-402, 2013.

Kim Y-H, Park J-W, Patel C, Kim D-Y. Polyethylene wear and osteolysis after cementless total hip arthroplasty with alumina-on-highly crosslinked polyethylene bearings in patients younger than thirty years of age. J Bone Joint Surg. 95: 1088-1093, 2013.

Lachiewicz PF, Soileau ES. Low early and late dislocation rates with 36- and 40-mm heads in patients at high risk for dislocation. Clin Orthop Rel Res. 471(2): 439-443, 2013.

Macias C, Bodugoz-Senturk H, Muratoglu OK. Quantification of PVA Hydrogel Dissolution in Water and Bovine Serum. Polymer 54(2):724-729, 2013.

Bedard NA, Callaghan JJ, Stefl MD, Willman TJ, Liu SS, Goetz DD. Fixation and wear with a contemporary acetabular component and cross-linked polyethylene at minimum 10-year follow-up. J Arthroplasty 29: 1961-1969, 2014.

Devane PA, Horne JG, Ashmore A, Mutimer J, Calvert G. A randomized prospective double-blind trial comparing X-linked with conventional polyethylene in THA. Minimum 10 Year Results. International Hip Society Closed Meeting, Rio de Janeiro, Brazil, 2014.

Australian Orthopaedic Association National Joint Replacement

Registry. Annual Report Hip and Knee Arthroplasty. Adelaide: AOA; 2014: 95, 2014.

Goel A, Lau EC, Ong KL, Berry DJ, Malkani AL. Dislocation rates following primary total hip arthroplasty have plateaued in the Medicare population. J Arthroplasty. Nov 26. pii: S0883-5403(14)00888-2. doi: 10.1016/j.arth.2014.11.012, 2014.

Snir N, Kaye ID, Klifto CS, et al. 10-year follow-up wear analysis of first-generation highly crosslinked polyethylene in primary total hip arthroplasty. J Arthroplasty 14; 29: 630-633, 2014.

Callary SA, Solomon LB, Holubowycz OT, Campbell DG, Munn Z, Howie DW. Wear of highly crosslinked polyethylene acetabular components: a review of RSA studies. Acta Orthopaedica 86(2), 159-168, 2014.

Bichara DB, Bodugoz-Senturk H, Ling D, Malchau H, Bragdon CR, Muratoglu OK. Osteochondral defect repair using a polyvinyl alcohol-polyacrylic acid (PVA-PAAc) hydrogel. Biomedical Materials 9 045012, 2014.

Goel A, Lau EC, Ong KL, Berry DJ, Malkani AL. Dislocation Rates Following Primary Total Hip Arthroplasty Have Plateaued in the Medicare Population. J Arthroplasty 30: 743-746, 2015.

Garvin KL, White TC, Dusad A, Hartman CW, Martell J. Low wear rates seen in THAs with highly crosslinked polyethylene at 9 to 14 years in patients younger than age 50 years. Clin Orthop Rel Res. 473: 3829-3835, 2015.

Oral E, Neils A, Muratoglu OK. High vitamin E content, impact resistant UHMWPE blend without loss of wear resistance. Journal of Biomedical Materials Research B 103 (4): 790-797, 2015.

Glyn-Jones S, Geraint ERT, Garfield-Roberts P et al. Highly crosslinked polyethylene in total hip arthroplasty decreases long-term wear: a double-blind randomized trial. Clin Orthop Rel Res. 473: 432-438, 2015.

Bragdon CR, Barr CJ, Nielsen CS et al. Minimum 10-year multi-center study of THR with highly cross-linked polyethylene and large diameter femoral heads. The Hip Society, Dallas TX, November 5-8, 2015.

Joyce TJ. Highly crosslinked polyethylene in total hip arthroplasty decreases long-term wear: A double-blind randomized trial. Clin Orthop Rel Res. 473: 439-440, 2015.

Goodman SB. Editorial Comment: 2014 Hip Society Proceedings. Clin Orthop Rel Res. 473: 430-431, 2015.

Lombardi AV, Berend KR, Morris MJ, Adams JB, Sneller MA. Large diameter metal-on- metal total hip arthroplasty: dislocation infrequent but survivorship poor. Clin Orthop Rel Res. 473: 509-520, 2015.

Nebergall AK, Troelsen A, Rubash HE, Malchau H, Rolfson O, Greene ME. Five-year experience of vitamin E-diffused highly cross-linked polyethylene wear in total hip arthroplasty assessed by radiostereometric analysis. J Arthroplasty June 31(6):1251-1255, 2016.

Nebergall AK, Rolfson O, Rubash HE, Malchau H, Troelsen A, Greene ME. Stable fixation of a cementless, proximally coated, double wedged, double tapered femoral stem in total hip arthroplasty: a 5-year radiostereometric analysis. J Arthroplasty. June 31(6): 1267-1274, 2016.

# CHAPTER 12

- - - - - - - - - - - - - - - - - -

# REDUCING THE INCIDENCE
# OF DISLOCATION OF THE HIP

DISLOCATION OF TOTAL HIP REPLACEMENTS has long been a constant, low level, disabling and very unsettling complication. As Charnley was driven to use smaller and smaller femoral heads because of rapid wear of the plastics, the risk of dislocation grew. His solution was to retain in continuity the integrity of the superior portion of the hip capsule with the osteotomized greater trochanter as a mandatory part of his surgical technique. This area of capsule served as a restraining thong, reducing the risk of dislocation. In his hands this technique was remarkably effective, but less so for many other surgeons. And it carried with it the special disadvantage of requiring the reattachment of the trochanter osteotomy. Because nonunion of the trochanter could lead to both limp and an increased risk of dislocation, it caused him to search many different avenues for a solution to get the greater trochanter to heal every time. His efforts in this regard were never fully successful.

The alternate solution that many others chose was to perform the operation without osteotomy of the trochanter. While this obviated the issue of nonunion of the greater trochanter, it caused other issues secondary to less extensive exposure. Despite this, the popularity of doing a THR without trochanteric osteotomy grew progressively until trochanteric osteotomy became far less common and was used primarily only for difficult situations.

The incidence of dislocation varies widely because it is so multifocal in etiology. Among the multiplicity of contributing factors are head size, degree of acetabular recess of the socket, the chamfer of the socket design, the attitude of the acetabular component, the attitude of the femoral component, leg length issues, neuromuscular diseases and the range of motion desired, achieved and used. The etiology of the initial hip pathology also plays an important role, exemplified by the higher rate in patients with developmental dysplasia. Age and gender are factors as well, with a higher risk in elderly patients, particularly elderly women after a hip fracture. The incidence of dislocation after revisions is clearly elevated and particularly those revisions of the acetabular component alone. The risk is singularly higher, obviously, if the trochanter fails to unite. Moreover, and independent of special features promoting dislocation, the incidence climbs slowly but relentlessly with time post operation. Wear can be an important factor in patients with conventional polyethylene, especially with a deeply penetrated small femoral head. Among the higher risks are patients who have dislocated before, and among the highest at risk are those who have had a revision for recurrent dislocation.

An exciting development with the promise of reducing the dislocation rate was the introduction of the extended lip liner. While the Harris Orthopaedic Laboratory had nothing to do with the creation of this innovation, we did participate in the critical evaluation of its potential advantages. Using our 3D range of motion model, we studied whether or not the extended lip liner, independent of its depth, extent or constraint, actually fulfilled its promise. Regrettably, it did not.

The salient observation was that the net effect of the extended lip liner is to tilt the face of the acetabular component, an effect that can be obtained with any acetabular component simply by adjusting its attitude. Despite these objective observations, the emotional appeal persists even today for its use.

The single most dramatic change in metal-on-polyethylene total hip surgery that effectively reduced the incidence of dislocation grew out of our work on creating a crosslinked polyethylene. The striking finding from our hip simulator work with crosslinked polyethylene was that we showed, quite separate from but in addition to the dramatic decrease in wear, the fact that no increase in wear existed between between using a 22 mm and a 32 mm head. This was in sharp contrast with conventional polyethylene. And it prompted the question – how far could we push this phenomenon? Our exploration of this question led us to study and subsequently support the use of large heads in a metal-on-plastic total hip arthroplasty when using crosslinked polyethylene. Even using 44mm heads, in the simulator we were unable to measure any wear of highly crosslinked polyethylene.

Our experimental work also showed that in fact it was possible, with the component positioned at proper attitude, to demonstrate that with a head size of 36 mm or more and an appropriate design of the articular recess into the plastic, the limiting factor which was now required for dislocation was actually <u>bone to bone</u> contact at the extremes of motion.

The subsequent step needed was to document this potential advantage by quantifying in patients the effect of the larger heads on the propensity for the hip to dislocate. That final step which established definitively the advantage of the big head in metal-on-polyethylene THA as the dominant factor in reducing dislocation was the prospective controlled randomized investigation of our polyethylene using 28 mm and 36 mm heads, done by Howie et al. Their results of this well-designed and excellent study showed a major and statistically significant reduction in the dislocation rate using the larger head diameter. The one-year incidence of dislocation using a 36 mm head was 0.8% versus the figure for a 28 mm head of 3.6%, a statistically significant reduction. Following revision operations, the parallel reduction using the 36 mm head versus a 28 mm head was from 12.2% among the 28 mm heads to 4.9%with the 36 mm heads. These are striking figures. Early (3-5 year) reports show no increase in wear rate but an increase in wear volume with 36 mm ID crosslinked sockets.

For both revision and primary THA the dual mobility acetabular component design incorporating crosslinked polyethylene has proven to be very effective in reducing the incidence of dislocation and appears to be cost-effective, providing an excellent approach to reduce this complication, based on data over the first decade or less.

With the dramatic reduction in the number of reoperations that were previously required because of the multiple types of failures caused by wear, lysis, granuloma formations, dislocation, fractures and component loosening and the the progressive increase in improved fixation, the profile of the common causes of revision changed after the introduction of crosslinked polyethylene and the wider use of ceramic-on-ceramic joints. And as a result of this surprising change, reoperations for recurrent dislocation rose to become among the most common. But, now the use of the larger head, whether in a ceramic-on-ceramic THA or metal-on-crosslinked polyethylene THA, and the widespread adoption of the dual mobility components not only in revision but in many primary operations, reduced the reoperation rate for recurrent dislocation drastically.

Another but transient contribution from the Harris Orthopaedic Lab to the problem caused by dislocation was the design of a constrained acetabular component. While this design of this constrained acetabular component had several important advantages, including the use of highly crosslinked polyethylene, the capacity to use larger femoral heads and preferential orientation of the available range of motion in the two directions most commonly associated with dislocation – namely flexion and internal rotation, or extension and external rotation – it was still beset by postoperative dislocation for some patients. This approach has largely been supplanted by the advantages of simply using larger head sizes or dual mobility components in the management of problems of recurrent dislocation.

Thus, an interplay of many aspects of our research impacted the problems of dislocation, the dislocation rate and valuable solutions for recurrent dislocation.

## SELECTED RELATED REFERENCES

Krushell RJ, Burke DW, Harris WH. Elevated Rim Acetabular Components: Effect on Range of Motion and Stability in Total Hip Arthroplasty. J. Arthroplasty Suppl 6 S53-S58, 1991.

Muratoglu OK, Bragdon CR, O'Connor DO, Jasty M, Harris WH. A Novel Method of Crosslinking Ultra-high Molecular Weight Polyethylene to Improve Wear, Reduce Oxidation, and Retain Mechanical Properties. Recipient of the HAP Paul Award. J. Arthroplasty 16(2):149-60, 2001.

Burroughs BR, Golladay GJ, Hallstrom B, Harris WH. A Novel Constrained Acetabular Liner Design with Increased Range of Motion. J. Arthroplasty 16(8) Suppl. 1:31-36, 2001.

Muratoglu OK, Bragdon CR, O'Connor DO, Jasty M, Harris WH. A Novel Method of Crosslinking Ultra-high Molecular Weight Polyethylene to Improve Wear, Reduce Oxidation, and Retain Mechanical Properties. Recipient of the HAP Paul Award. J. Arthroplasty 16(2):149-60, 2001.

Muratoglu OK, Bragdon CR, O'Connor DO, Perinchief RS II, Estok DM, Jasty M, Harris WH. Larger Diameter Femoral Heads used in Conjunction with a Highly Crosslinked Ultra-High Molecular Weight Polyethylene: A New Concept. J. Arthroplasty 16(8) Suppl. 1: 24-30, 2001.

Burroughs BR, Golladay GJ, Hallstrom B, Harris WH. A Novel Constrained Acetabular Liner Design with Increased Range of Motion. J. Arthroplasty 16(8) Suppl. 1:31-36, 2001.

Muratoglu OK, Bragdon CR, O'Connor DO, Perinchief RS II, Estok DM, Jasty M, Harris WH. Larger Diameter Femoral Heads used in Conjunction with a Highly Crosslinked Ultra-High Molecular Weight Polyethylene: A New Concept. J. Arthroplasty 16(8) Suppl. 1:24-30, 2001.

Burroughs BR, Rubash HE, Harris WH. Femoral head sizes larger than 32 mm against highly crosslinked polyethylene. Clin. Orthop Rel Res. 405:150-157, 2002.

Burroughs BR, Hallstrom B, Golladay GJ, Hoeffel D, Harris

WH. Range of Motion and Stability with 28-, 32-, 38-, and 44-mm Femoral Head Sizes: An In Vitro Study. J Arthroplasty (20)1:11-19, 2005

Bragdon CR, Jasty M, Muratoglu OK, O'Connor DO, Harris WH. Third-body wear of highly crosslinked polyethylene in a hip simulator. J. Arthroplasty, 18(5):553-561, 2003.

Muratoglu OK, Greenbaum ES, Bragdon CR, Jasty M, Freiberg AA, Harris WH. Surface analysis of early retrieved acetabular polyethylene liners: a comparison of conventional and highly crosslinked polyethylenes. J. Arthroplasty 19(1): 68-77, 2004.

Burroughs BR, Hallstrom B, Golladay GJ, Hoeffel D, Harris WH. Range of Motion and Stability with 28-, 32-, 38-, and 44-mm Femoral Head Sizes: An In Vitro Study. J. Arthroplasty (20)1;11-19, 2005.

Bragdon CR, Jasty M, Muratoglu OK, Harris WH. Third-body wear testing of a highly cross-linked acetabular liner: the effect of large femoral head size in the presence of particulate poly(methyl-methacrylate) debris. J. Arthroplasty. April;20(3):379-85, 2005.

Bragdon CR, Greene ME, Freiberg AA, Harris WH, Malchau H. Radiostereometric analysis comparison of wear of highly cross-linked polyethylene against 36- vs 28-mm femoral heads. J. Arthroplasty. 22(6 Suppl 2):125-129, 2007.

Estok DM 2nd, Burroughs BR, Muratoglu OK, Harris WH. Comparison of hip simulator wear of 2 different highly cross-linked ultra high molecular weight polyethylene acetabular components using both 32- and 38-mm femoral heads. J. Arthroplasty. Jun 22(4) 581-589, 2007.

Plank GR, Estok DM 2nd, Muratoglu OK, O'Connor DO, Burroughs BR, Harris WH. Contact stress assessment of conventional and highly crosslinked ultra high molecular weight polyethylene acetabular liners with finite element analysis and pressure sensitive film. Journal of Biomedical Materials Research. Part B, Applied Biomaterials. Jan.;80(1):1-10, 2007.

Howie DW, Holubowicz OT, Middleton R. Large femoral heads decrease the incidence of dislocation after total hip arthroplasty: a randomized controlled trial. J Bone Joint Surg Am Jun 20;94(12):1095-1102, 2012.

Bragdon C, Doerner M, Rubash H, Kwon YM, Martell J, Clohisy J, White R, Della Valle C, Berry D, Jarrett B, Lachiewicz P, Bertin K, Johansson P, Palm H, Malchau H, Harris WH. Clinical Multi-centric Studies of the Wear Performance of Highly Crosslinked Remelted Polyethylene in THR. Clin Orthop Rel Res 471(2): 393-402, 2013.

Estok DM 2nd, Burroughs BR, Muratoglu OK, Harris WH. Comparison of hip simulator wear of 2 different highly cross-linked ultra high molecular weight polyethylene acetabular components using both 32- and 38-mm femoral heads. J. Arthroplasty Jun 22(4):581-589, 2007.

Callary SA, Field JR, Campbell DG. The rate of wear of second-generation highly crosslinked polyethylene liners five years postoperatively does not increase if large femoral heads are used. Bone Joint J Dec;98-B(12):1604-1610, 2016.

Howie DW, Hołubowicz OT, Callery SA. The wear rate of highly cross-linked polyethylene in total hip replacement is not increased by large articulations: A randomized controlled trial. J Bone Joint Surg Am. Nov 2;98(21):1786-1793.

Gonzalez AI, Bartolone P, Lubbeke A, Dupuis LE, Peter R, Hoffmeyer P, Christofilopoulos P. Comparison of dual-mobility cup and unipolar cup for prevention of dislocation after revision total hip arthroplasty. Acta Orthop Feb;88(1):18-23, 2017.

Martino I, D'Apolito R, Soranoglou VG, Poultsides LA, Sculco TP. Dislocation following total hip arthroplasty using dual mobility acetabular components: a systematic review. Bone Joint J Jan;99-B(1 Suppl A):18-24, 2017.

Jauregui JJ, Pierce TP, Elmallah RK, Cherian JJ, Delanois RE, Mont MA. Dual mobility cups: an effective prosthesis in revision total hip arthroplasties for preventing dislocations. Hip Int. Jan-Feb;26(1):57-61, 2016.

Epinette JA, Harwin SF, Rowan FE, Tracol P, Mont MA, Chughtai M, Westrich GH. Early experience with dual mobility acetabular systems featuring highly cross-linked polyethylene liners for primary hip arthroplasty in patients under fifty five years of age: an international multi-centre preliminary study. Int Orthop Mar;41(3):543-550, 2017.

Epinette JA, Lafuma A, Robert J, Doz M. Cost-effectiveness model comparing dual-mobility to fixed-bearing designs for total hip replacements in France. Orthop Traumatol Surg Res. Apr;102(2):143-148, 2016.

# CHAPTER 13

- - - - - - - - - - - - - - - - - -

# QUANTIFICATION

QUANTIFICATION OF IMPORTANT FEATURES OF every surgical endeavor plays a key role in communication, education, and advancement in surgery. The Harris Orthopaedic Lab made important contributions to critical quantification of key features of total hip surgery in two areas. The first was the major, site-specific rating of hip functions called the Harris Hip Score (HHS). The second was a focused assessment of the efficacy of the quality of the use of cement for cement fixation for the femoral component. Both had worldwide utility and acceptance. So too did our method of quantification of the diagnosis of whether or not a femoral compound was solidly fixed to the femur.

Curiously, while the Harris Hip Score has played a major role in the evaluations of nearly all aspects of the development of new devices and innovative techniques of THA, its origin antedated the introduction of THA to the United States. Specifically, from the beginning of this new standard of evaluation of hip function, its scope encompassed and exceeded that of any single hip operation or its complications.

Dissatisfaction lay at the basis of its creation. The dissatisfaction came from the inherent deficiencies extant in the existing vehicles available for us to critique our efforts as hip surgeons. Several systems were aimed to evaluate the efficacy of operation X but did not address the truly important criteria of the aggregate status of the individual hip preoperatively versus how this changed postoperatively, independent of the type of operation done! For example, in some rating systems assessing cup arthroplasty, motion was rated so highly that other aspects of hip function, such as pain relief and function of the patient, which should have been valued more heavily, were downgraded. Additional features such as the amount of support used or the distance walked were often ignored and limp was not quantified.

With four operations in use at the inception of the Harris Hip Score, cup arthroplasty, hip fusion, osteotomy of the hip and total hip arthroplasty, the key concept behind the HHS system was to quantify, as accurately as possible among the many semi-quantifiable elements involved, those special issues most important to the patient, not to those specifically valuable for a given operation. And these were to be contrasted before and after surgery. It addressed how accurately we could assess the status of the patient before and after surgery and even simply over the passage of time in terms of which issues of hip function were most valuable to the patient. For these reasons, pain relief and functional capacity took center stage. Yes, motion could be important, but its value lay primarily in its role in increasing or limiting function.

A uniquely valuable feature of the quantification in this system of evaluation was the numerical quantification of each element in the method by the assignment of points to each segment of the evaluation, i.e., individual numerical scores for various degrees of pain or for distance walked. The special advantage that grows from this feature is the ability for each assessment of the patient to be assembled into a simple, crucial total assessment, a single overall score for that patient.

Critics argued against a single overall score, preferring simply to compare one group of patients against another, not by a total score per individual patient, but rather by percentage of patients with, for example, severe pain or the need for a cane. The fallacy of that argument is quickly revealed by a single example. Suppose that within 100 post-operative patients, 30% used two crutches, 30% had severe pain and 30% could not walk beyond three blocks. While this information may be of some limited value when compared to another group of patients or that same group preoperatively versus postoperatively, the absence of a single overall score for each patient makes it impossible to distinguish within that data set between one extreme profile of those 100 patients which represents 30 severely disabled patients all of whom had each of these three bad results or the other extreme profile consisting of 90 patients each of whom had only one each of these three bad results. Obviously, other alternative distributions of the patients also exist, consisting of some groupings of these 100 same patients somewhere between the two extremes of the 30 or the 90 patients described. To reemphasize, a single overall score per each individual patient is mandatory for any representative quantification of the status of the patient at any evaluation.

Moreover, the single overall score per patient makes it easy to quantify results per individual patient, as well as also per group. Also, the magnitude of the progress or regress that results from the surgical intervention or from the passage of time can be readily quantified for each patient.

An understanding of the nature of some of the compromises required to be able to have any system of evaluation of the hip is important. For example, a complex relationship exists between the distance a patient can walk versus the amount of support required to walk that distance. If a patient could walk well over long distances using a cane, that is quite different from being "cane-free but limping badly". This is why questions such as distances walked require the concomitant question detailing the support needed to walk that distance.

The publication of the Harris Hip Score occurred in 1969 in an article dealing with cup arthroplasty for dislocation or traumatic arthritis of the hip. But the unique conceptual features of this system, and particularly its quantification of the specific overall numerical result for each patient at each time, led to its wide and prolonged acceptance (now 48 years) as the key, site-specific form of quantification of the outcomes of various types of hip surgery.

The utility of this concept and specifically this rating system is well reflected by its remarkable rate of citations. By a huge margin, scientific citation of this article is the single leading article cited by others, not only in all of hip and knee arthroplasty in the past 50 years, but it is also the leading article among all citations of any sort dealing with orthopaedic surgery and also with orthopaedic hip research.

In fact, in the ranking of the 50 highest cited papers in the report on citations in the area of hip and knee arthroplasty by Holzer and Holzer, they found that not only was this paper describing the Harris Hip Score the highest citation of any, by a wide margin, but also that nine of the top 50 citations in the field of hip and knee arthroplasty arose from the Harris Orthopaedic Laboratory.

Another important contribution to quantification dealt with evaluation of femoral cementing techniques. It, too, grew out of the desire to be able to quantify the efficacy of various "improvements" in cementing techniques discussed in the chapter on fixation (Chapter 8). By defining selected findings of the Roentgen appearance of cemented femoral components, not only could these features be related to both the cemented technique used and the long-term outcome, but most importantly it could provide both short-term and long-term feedback to the individual surgeon. For example, the experimental evidence strongly favored the plugging of the femoral canal, but it became essential to know if that technique made a difference in the longevity of cement fixation in real life. Equally so, the use of a cement gun, centrifugation or other porosity-reduction techniques and pressurization of the cement were all quantified using these systems, and these changes could be evaluated in terms of outcome independent of the method of use of the techniques. It was in order to present these important assessments that we developed the criteria for evaluating the quality of femoral cementing and thus the capacity to determine their true utility.

All of the assessments of variations in cementing techniques were also ultimately related to our widely used quantification of whether or not a given cemented femoral component was loose, using a constellation of radiographic appearances with which to judge gradations in fixation as definite, possible, or probably loosening. This quantification method contributed strongly to the development of improving cementing techniques.

## SELECTED RELATED REFERENCES

Shepherd MM. Assessment of Functions after Arthroplasty of the Hip. J Bone Joint Surg 36-B:354-363, 1954.

Shepherd MM. A Further Review of Results of Operations in the Hip Joint. J Bone Joint Surg 42-B:177-204, 1960.

Larson CB. Rating Scale for Hip Disabilities. Clin Ortho Rel Res. 31:85-93, 1963.

Harris WH. Traumatic Arthritis of the Hip after Dislocation and Acetabular Fractures: Treatment by Mold Arthroplasty. An End-Result Study Using a New Method of Result Evaluation. J Bone Joint Surg., 51-A:737-755, 1969.

Harris WH. Loosening. In: The Hip, Proceedings of the Sixth Open Scientific Meeting of the Hip Society. Ed. Carl S Nelson. St. Louis, CV Mosby Co., pp. 162-175, 1978.

Harris WH Jr, McCarthy JC, O'Neill DA. Loosening of the Femoral Component of Total Hip Replacement after Plugging the Femoral Canal. In: The Hip, Proceedings of the 10th Open Scientific Meeting of the Hip Society, St. Louis: CV Mosby Co., pp. 228-238, 1982.

Harris WH, McCarthy JC, O'Neill DA. Femoral Component Loosening Using Contemporary Techniques of Femoral Cement Fixation. J. Bone Joint Surg., 64-A: 1063-1067, 1982.

Oh I, Bourne RB, Harris WH. The Femoral Cement Compactor. An Improvement in Cementing Technique in Total Hip Replacement. J. Bone Joint Surg., 65-A:1335-1338, 1983.

Burke DW, Gates EI, Harris WH. Centrifugation as a Method

of Improving Tensile and Fatigue Properties of Acrylic Bone Cement. J. Bone Joint Surg., 66-A:1265-1273, 1984.

Harris WH, McGann WA. Loosening of the Femoral Component after Use of the Medullary-Plug Cementing Technique. Follow-up Note with a Minimum Five-year Follow-up. J Bone Joint Surg., 68-A:1064-1066, 1986.

Davies JP, O'Connor DO, Burke DW, Jasty M, Harris WH. The Effect of Centrifugation on the Fatigue Life of Bone Cement in the Presence of Surface Irregularities. Clin Orthop Rel Res 229:156-161, 1988.

Rubash HE, Harris WH. Revision of Nonseptic, Loose, Cemented Femoral Components Using Modern Cementing Techniques. J. Arthroplasty 3:241-248, 1988.

Mulroy RD Jr, Harris WH. The Effect of Improved Cementing Techniques on Component Loosening in Total Hip Replacement. J Bone Joint Surg. 72-B:757-760, 1990.

Mulroy RD Jr, Sedlacek RC, O'Connor DO II, Estok DM, Harris WH. Technique to Detect Migration of Femoral Components of Total Hip Arthroplasties on Conventional Radiographs. J. Arthroplasty 6, Suppl., S1-S4, 1991.

Barrack RL Jr, Mulroy RD, Harris WH. Improved Cementing Techniques and Femoral Component Loosening in Young Patients with Hip Arthroplasty: a 12 year radiographic review. J Bone Joint Surg. 74-B: 385-389, 1992.

Harris WH. Total Hip Replacement in the Middle-Aged Patient. Contemporary Cementing for Fixation of the Femoral Component. Orthop Clin N Amer. 24:611-616, 1993.

McLaughlin R, Harris WH. A Composite Plug for Occluding the Femoral Canal Prior to Cementing a Total Hip Femoral Component. Orthopaedic Review 23:344-356, 1994.

Harris WH. The case for cementing all femoral components in total hip replacement. Canad. J. Surg., 38: Suppl 1, S55-60, Feb., 1995.

Pierson JL, Harris WH. Effect of improved cementing techniques on the longevity of fixation in revision cemented femoral arthroplasties: Average 8.8 year follow-up period. J. Arthroplasty 10:581-591, 1995.

Harris WH. Improved Long-Term Results of Femoral Fixation in Revision Surgery Using Modern Cementing. Ch. 31. 289-294. In: Total Hip Revision Surgery. Galante JO, Rosenberg AG, Callaghan JJ, Eds. Bristol-Myers, 1995.

Mulroy WF II, Estok DM, Harris WH. Total hip arthroplasty with use of so-called second-generation cementing techniques. A fifteen-year-average follow-up study. J. Bone Joint Surg. 77-A (12) 1845-1852, 1995.

Freiberg AA, Harris WH. The femoral stent: Surgical technique for a new method of cementing the femoral component in the presence of a large cortical defect. J. Orthop Rel Res 24:501-503, 1995.

Mulroy WF, Harris WH. Revision total hip arthroplasty with use of so-called second-generation cementing techniques for aseptic loosening of the femoral component. A fifteen year follow-up study. J Bone Joint Surg. 78-A:325-330, 1996.

Smith S II, Estok DM, Harris WH. Total Hip Arthroplasty with Use of Second- Generation Cementing Techniques: An Eighteen-Year-Average Follow-up Study. J Bone Joint Surg. 80-A (11):1632-1640, 1998.

Bourne RB, Rorabeck CH, Skutek M, Mikkelsen S, Winemaker M, Robertson D. The Harris Design-2 total hip replacement fixed with so-called second-generation cementing techniques. A ten to fifteen-year follow-up. J Bone Joint Surg 80(12):1775-1780, 1998.

Smith SE II, Estok DM, Harris WH. 20-Year Experience with Cemented Primary and Conversion Total Hip Arthroplasty Using So-Called Second-Generation Cementing Techniques in Patients Aged 50 Years of Younger. J Arthroplasty 15:3, 263-273, 2000.

Herberts P, Malchau H. Long-term registration has improved the quality of hip replacement: a review of the Swedish THR Register comparing 160,000 cases. Acta Orthop Scand 71(2):111-121, 2000.

Söderman P, Malchau H. Is the Harris Hip Score system useful to study the outcome of total hip replacements? Clin Orthop Rel Res;384:189-197, 2001.

Hauser DL, Wessinger SJ, Condon RT, Golladay GJ, Hoeffel DP, Gillis DJ, Merrill DR, Chaisson D, Freiberg AA, Estok DM, Rubash HE, Malchau H, Harris WH. An Electronic Database for Outcome Studies That Includes Digital Radiographs. J. Arthroplasty 16(8) Suppl. 1, 71-75, 2001.

Behairy Y, Harris WH. Mode of Loosening of Matte-Finished Femoral Stems in Primary Total Hip Replacement. Saudi Med J. 23:10, 1187-94, 2002.

Bragdon CR, Estok DM, Malchau H, Karrholm J, Yuan X, Bourne R,Veldhoven J, Harris WH. Comparison of two digital radiostereometric analysis methods in the determination of femoral head penetration in a total hip replacement phantom. J. Orthop Rel Res. 22(3); 659-64, 2004.

Bragdon CR, Estok DM, Malchau H, Karrholm J, Yuan X, Bourne R, Veldhoven J, Harris WH. Comparison of two digital radiostereometric analysis methods in the determination of femoral head penetration in a total hip replacement phantom. J. Orthop Rel Res. 22(3); 659-64, 2004.

Smith A, Fagnani M, Condon R, Wessinger SJ, Dorrwachter J, Bragdon C, Malchau H. Web-Accessible Orthopaedic Outcomes Engine: The Harris Joint Registry at MGH. The Orthopaedic Journal at Harvard Medical School. 7:129-132, 2005.

Bragdon CR, Martell JM II, Estok DM, Greene ME, Malchau H, Harris WH. A new approach for the Martell 3-D method of measuring polyethylene wear without requiring the cross-table lateral films. J. Orthopaedic Research August 23 (4): 720-725, 2005.

Bragdon CR, Martell JM, Greene ME, Estok DM, Thanner J, Karrholm J, Harris WH, Malchau H. Comparison of Femoral Head Penetration Using RSA and the Martell Method. Clin Orthop. 448: p. 52-57: 2006.

Ioppolo J, Borlin N, Bragdon C, Li M, Price R, Wood D, Malchau H, Nivbrant B. Validation of a low-dose hybrid RSA and fluoroscopy technique: Determination of accuracy, bias and precision. Journal of Biomechanics.40(3):686-92, 2007.

Bragdon CR, Greene ME, Freiberg AA, Harris WH, Malchau

H. Radiostereometric analysis comparison of wear of highly cross-linked polyethylene against 36- vs 28-mm femoral heads. J. Arthroplasty Sept. 22(6 Suppl 2):125-9, 2007.

Kelly JC, Glynn RW, O'Briain DE, Felle P, McCabe JP. The 100 classic papers of orthopaedic surgery: a bibliometric analysis. Bone and Joint Journal doi:10.1302/0301-620X.92B10.24867, September 2010.

Lefaivre KA, Shadgan B, O'Brien PJ. 100 most cited articles in orthopaedic surgery. Clin Orthop Relat Res;469(5):1487-1497, May 2011.

Holzer LA, Holzer G. The 50 highest cited papers in hip and knee arthroplasty. J Arthroplasty Mar 29(3):453-457, 2014.

Ahmad SS, Albers CE, Buchler L, Kohl S, Ahmad SS, Klenke F, Siebenrock KA, Beck M. The hundred most cited publications in orthopaedic hip research - a bibliometric analysis. Hip International 26(2)199-208, 2016.

# CHAPTER 14

- - - - - - - - - - - - - - - -

# SYSTEMS APPROACHES

AT TWO DISTINCT POINTS IN the evolution of total hip surgery I introduced the superimposition of a <u>systems</u> <u>approach</u> on the organization of total hip implants. Each system played a crucial role, roles now easily dismissed as obvious, but which at their time were were important and even "radical".

The first dealt with the burning issue at the time, the size of the femoral head. The second dealt with the raging conflict between cement fixation and cementless fixation. Both of these major changes in how implants were designed and organized had their origin in the Harris Orthopaedic Lab and have now become worldwide, even ubiquitous.

A defining characteristic of total hip implants during the early decades of their development was the diameter of the femoral head. As Charnley experimented with his revolutionary concept of a metal-on-plastic total hip, the Teflon-like plastic was his initial and singularly unsuccessful iteration. The rapid wear of this plastic forced him to abandon his starting proposal for the optimum femoral head diameter, namely 41.5 mm. Forced by the alarming, rapid wear of the plastic, he progressively reduced the diameter of the femoral head, finally settling on 22.5 mm. The aim was that the smaller head both reduced friction, hence the "low friction arthroplasty", and increased the thickness of the available plastic for any given size of the acetabular component. This greater thickness of the plastic extended the longevity of the construct. Subsequently, even after he had abandoned the Teflon-like plastic and had demonstrated the advantages of the ultra high molecular weight polyethylene, he continued his reliance on the very small head diameter in this new metal-on-polyethylene articulation.

During the decade of the '60s, the first break against this small head diameter concept arose with the introduction by Mueller of the Charnley-Mueller THR, with Mueller's recommendation of a 32 mm femoral head diameter. A key reason supporting this larger head size was the reduced risk of dislocation compared to a 22 mm head. Although many factors contribute to any dislocation, such as the specifics of the chamfer of the acetabular component, the position of the acetabular component, the anteversion of the femoral component, neurologic diseases etc., it is also true that a larger head diameter – all other things being equal – permitted a greater range of motion with less risk of dislocation. (See Chapter 12, on dislocation)

The frictional torque of the larger head was clearly greater, and the distance traveled by any given point on the femoral head through the same arc of motion was greater for the larger head. Whether these factors which probably accelerated wear were offset by the reduced load per unit area afforded by the larger head remained unknown at the time, and thus controversial. The multiple complexities of measuring actual polyethylene wear in use in humans made it impossible to quantify exactly the net outcome of all these factors on wear during those early days. These complexities also rendered uncertain whether the early hip simulators or pendulum comparators actually replicated the human use conditions sufficiently accurately to be truly representative.

Thus it was that two strongly held positions developed, Charnley's 22 mm head diameter and Mueller's 32 mm head diameter. Robust arguments (more heat than light) supported each, in the absence of any definitive data.

When two new designs appeared from the Western Hemisphere,

they both adopted an intermediate head size. The T-28 THA from Amstutz adopted the 28 mm head diameter and the Harris THA favored the 26 mm head diameter. These choices were based on the election of an intermediate size between 22 and 32 mm in the absence of any definitive data for resolution of the true optimum. These designers hoped to gain partial advantage of the positive aspects of each extreme head size.

Champions for each system advanced the conceptual rationale for each head size vigorously and thus four distinct, competing camps developed worldwide, all four actively defended and advocated. For most surgeons all the other design features of the various systems were subservient in the decision-making process about the choice of head diameter. Four totally incompatible camps developed, rigidly advocating each femoral head diameter.

The sudden dissolution of these four distinct silos resulted from the introduction of my HD-2 total hip system, which incorporated the option of any of these four widely used head diameters. Central to the creation of this inclusionary concept was the insight that the selection of a given hip system purely on the basis of head diameter alone forced the surgeons to abandon all the other advantages of any given stem design. That meant, for example, that if a surgeon chose the 32 mm head diameter, that surgeon was forced to accept all the specific features of the Charnley-Mueller design, and not only for the femoral component, but also for the acetabular component features, regardless of their advantages or disadvantages.

Independent of the head size issue, the femoral component of the HD-2 implant had numerous advantages such as its broad medial border, a rounded medial border, absence of sharp edges, and the I-beam configuration. The acetabular component, regardless of the ID of the acetabular recess, had replaceability of the polyethylene liner via the metal-backing feature of the acetabular design, a specific chamfer design and an advanced acetabular component inserter which facilitated optimization of the positioning of the acetabular component. Prior to introducing a system of multiple head diameters, all these features would be lost unless the surgeon selected the 26 mm head, and vice versa for the Charnley 22 mm design, the 32 mm Charnley-Mueller system and the 28 mm Amstutz system.

By taking the startling and unprecedented step of including all four head sizes but retaining each of the other features of the HD-2 package, this systems approach made the full advantages of the HD-2 design available to any surgeon who wished to avail himself or herself of those features, independent of head size.

In retrospect, this would appear both simple and self-evident. To understand the full impact of this striking change at that time, one must relive those intense debates and strong feelings extant at that time totally in support of one, but only one, choice on the dominating issue of head size.

In fact, having all four head sizes in one system was considered to be a serious challenge to this inclusive concept, suggesting that doing so would undermine the integrity the Harris system since "he advocates 26 mm heads."

The advantage of having multiple femoral head diameters in any one THA design is obvious. It allows the surgeon to choose the <u>other</u> key design features, not being dominated by head size to choose but one design. Thus the options for the patient were greatly enlarged by superimposing this systems aspect of total hip designs.

The second major and even greater innovation contributing to the concept of a "system design" for total hip implants occurred with the introduction of the "Total System" in 1983. During the ascendency of the erroneous, widespread belief that all periprosthetic osteolysis was attributed to particulate bone cement. i.e., lysis was "cement disease" (See Chapter 11), vigorous debate raged about whether fixation of components should be done only with bone cement or totally without bone cement. As a result, all THA designers and implant manufacturers focused on completely separate, unrelated programs for either a cementless or a cemented set of components and implants. This was universal.

The disadvantages to the surgeon, to the OR personnel, to the hospital and thus to the patient, were large. Much of the huge inventory of instruments and implant components were essentially duplicated.

Storage requirements were increased. Familiarity with the many different implants, instruments and techniques was compromised and cost was increased. Often several different additional vendors were involved in planning for a single case.

In creating The Total System, the first integrated total hip system which consisted of <u>both</u> cemented and cementless total hip implants and the instruments and techniques, these multiple disadvantages were eliminated. For example, all the instruments that served the same function for cementless and cemented operations, such as for acetabular preparation, leg length measurements, and even femoral, the rasp handles were identical and could be used for either operation.

Both the set-up for the operation as planned and the necessary changes to accommodate any unexpected alternative were substantially simplified. And usually only one vendor was involved. This was often true including those infrequent but critical examples of suddenly requiring an implant or instrument not instantly available within the hospital itself.

Particularly useful was the unified system during the time where the hybrid THR concept became widespread. Similarly, during the interval when it became clear that a selected hemispherical cementless acetabular component had major advantages for long-term successful fixation but that certain cementless femoral implants were less satisfactory, combining a cementless acetabular component with a cemented femoral component (the hybrid approach) commanded wide use. To be able to do hybrid THR reconstruction from one vendor incorporating all aspects of a single system was far more familiar, easier, faster, and cheaper.

The many advantages of having an integrated system of cemented and cementless implants proved to be so valuable at so many levels that this concept was widely adopted throughout all companies and around the world. It is now a universal standard.

# CHAPTER 15

------------------------------------------------

# TOTAL JOINT EDUCATION COURSES

BECAUSE POSTGRADUATE MEDICAL EDUCATION IS so vital, it needs many avenues. The rate of change in medicine is massive but the acceleration of that rate is even more so, as is emphasized throughout contemporary life by Tom Friedman's book "Thank You for Being Late". For over four and half decades the Harris Orthopaedic Laboratory played a key role in one form of postgraduate surgical education, the short, multi-day, focused continuing education course. The origin of and the characteristics of this course on total joint surgery merit description.

To understand this process, now so seemingly natural, the time-machine must reverse 50 years, back to the mid-Sixties. Although Charnley introduced metal-on-plastic THR using conventional polyethylene in 1962, still by 1970 total hip replacement (THR) was not permitted in the USA. The use of polymethyl methacrylate (bone cement) had not been permitted by the FDA. As a preliminary step to introducing THR to this country, 100 selected surgeons were granted the right to receive and use polymethyl methacrylate, so as to begin to generate US data on this strange, new operation. Massive, vocal opposition to this selective designation from the general body of non-selected surgeons doing hip surgery was predictable and voracious.

Still, contemplation of a general release of this material and thus the general release of this type of revolutionary and vastly different surgery generated most serious and compelling reservations. With only few exceptions, no one in the USA had seen the surgery, understood the chemistry and toxicology of polymethyl methacrylate, been taught or had experience with this surgery, or was well educated on the totality of everything from patient selection to technical details to complications or outcomes. The revision of failed total hip replacements was yet another unknown, of an even greater complexity, ignorance and risk.

How do you suddenly – if polymethyl methacrylate were released – simultaneously teach thousands of surgeons a complex, unique surgical procedure that they have never seen while simultaneously protecting the citizenry?

The resolution called for a massive, nationwide, comprehensive crash course in THR surgery, in every aspect. To do this the American Academy of Orthopaedic Surgeons chose four surgeons with experience in this operation to create a comprehensive packet of training videos which could serve as the heart and soul of the needed massive educational venture. These training videos were to be displayed as post-graduate education at multiple sites across the nation. Needless to say, a nationwide educational venture of this scale and intensity had had never been undertaken before. The four surgeons selected were Mark Coventry of the Mayo Clinic, Phil Wilson, Jr. of the Hospital for Special Surgery, Harlan Amstutz of UCLA and me. The locus for generating the videos was carefully chosen, the film school of UCLA. The charge to this small band was to educate all USA surgeons doing reconstructive hip surgery, as well as those wishing to do so, in every facet of indications, risks, complications, surgical approaches, device techniques, options, use of polymethyl methacrylate and its pitfalls, radiographic interpretation, reoperations including indication, types, complexities, precautions, execution, and outcomes.

The week of videorecording by all four of us was more than full, primarily with lectures and surgery on cadavers. It was not only grueling, exhausting, and demanding but also comprehensive and effective.

Then came information transfer. For this, multiple presentations were made and made repeatedly across the country. The original display of my presentations in New England was made at the Shriners' Auditorium of the Burns Unit in Boston in the fall of 1970, with knowledgeable speakers contributing live presentations to augment the video material. Because the demand was so large, additional performances were done there and at multiple sites across the land.

Interestingly, it was this unique set of circumstances that created my first interaction with the Department of Continuing Education at Harvard Medical School, who provided the administrative and educational oversight of this initial venture. The overwhelming demand coupled with the enthusiastic response prompted a repeat course the following fall. Thus was born an annual 3-4 day postgraduate educational exercise each fall dealing with total hip replacement, and eventually total knee replacement, overseen by the Department of Continuing Education at Harvard Medical School. Its longevity was 49 years, save one year, one of the longest continuing education courses in the history of the Medical School.

Because of the remarkable durability of this course, it is valuable to describe its characteristics and guiding principles. A general description of the format of the courses would not distinguish it from many other similar ventures. Lectures predominated, interspersed with videos, exhibits, live surgery, discussions and debates. Rather, it was the principles behind the selection of both the speakers and the subjects that gave it its unique recognition.

Beyond the absolute requirement of being objective and noncommercial, the defining characteristics were:

- Scientific rigor and overt integrity
- A full recitation of all major implants and techniques demonstrating the full spectrum of joint surgery as practised today, but also with....

- A strongly articulated vision of the direction the field is heading for tomorrow
- Marked emphasis on data and quantification, versus unsupported enthusiasm and anecdotes
- Outreach to the creative, disciplined contributors to the field regardless of their recognition, location or age
- The commitment that if a person had only one talk to give but it was really important and no one else had those data, they too were invited.

As the field grew and the course grew, so did the faculty. Notable among the foreign participants were Charnley, Mueller, Ganz, Wagner, Freeman, Buckholz, Giacometti, Herberts, Malchau, Letournel, Hart, Porter and many others, along with all the major North American contributors to progress in total joint surgery of the past 50 years.

As is true in all such endeavors, we, the faculty, learned the most. And we were enriched the most by the interchange. This was our reward for the hard work of creating it.

Dan Berry, Past President of the American Academy of Orthopaedic Surgeons (AAOS) and emeritus Chief of Orthopaedics at the Mayo Clinic, repeatedly said that he felt this course was the finest in the land. He always was delighted to participate because it gave him the best window on the field as it currently existed plus the optimum view of the most likely future directions.

# CHAPTER 16

- - - - - - - - - - - - - - - - -

# THE HARRIS REGISTRY

As the Harris Registry at the MGH joined forces in 2017 with three select registries throughout the world to receive the third Kappa Delta Award granted to the Harris Orthopaedic Lab for outstanding science contribution in Orthopaedic Surgery, specifically for the article entitled "Arthroplasty Implant Registries over the Past Five Decades", it is meaningful to reflect on the origin and development of the Harris Registry. As noted in the dedication of this volume, "disciplined dissatisfaction" is a major motivating factor for scientific advance. That specific dissatisfaction which generated the Harris Registry was the same one that fostered the Harris Hip Score, namely the lack of any method that would permit quantitative assessment of the results of hip surgery. Concomitant with the permission from the FDA to permit total hip surgery to be done in the United States in 1969 and the publication and subsequent wide adaption of the Harris Hip Score, also in 1969, it was evident that detailed quantification of outcomes of many aspects of hip surgery needed to be on a broad base. Not only must this apply to evaluation of overall results, but granular assessments would be required to evaluate critically even the more focused aspects such as novel implant designs, materials changes and alterations in technique. Another essential feature was the need to quantify the effects on the passage of time, to evaluate what worked and what did not over the long term. Thus began the Harris Registry. And it started with the Harris Hip Score.

Many important registries preceded it, many others augmented it and many others exceeded its scope. Nevertheless, it has and will contribute substantially. Among those registries extant at its inception were the excellent Mayo Clinic system established by Mark Coventry, the detailed data collection by John Charnley at Wrightington, England, that of D'Aubigne and Postel in Paris, and others.

Initially the data collected followed those items needed for fulfilling the Harris Hip Score, but that set rapidly expanded in several diverse directions, first to evaluate additional issues associated with the increasing revision surgery and then broadly, to include the important but often clinically silent radiographic changes that could portend subsequent or impending failure.

Such detailed interpretive assessment is manpower-intensive. Initially, individual paper questionnaires were generated to command inclusive catchment of these specific observations, both per patient, per medical or surgical interaction per patient, per type of surgical episode, and all, over time. Central to the generation of uniform data was the detailed weekly review by the director of the lab of all data admitted into the registry. In essence the Arthroplasty Fellows bore the major first line responsibility in preparing and presenting in detail all data to the Director of the lab at a weekly intake meeting, a requirement for information to enter the system.

The history of the technical features of the Registry reflects advances in data collection over the past 50 years. It began with the IBM punch card system. Fortunately the digital age was awakening, and the Registry passed through multiple stages beginning with the Commodore system, to the Vax, through multiple subsequent iterations to reach the level of 4-D and then into more massive capabilities including Remedy, and currently Epic.

Similarly, expansion of the scope of patient inclusion morphed from only my cases to a broader inclusion of multiple hip surgeons at the MGH, to data collected from the broad sweep of all patients in the Partners HealthCare system and then to a broad array of institutions within North America and finally multiple studies drawing data from many registries around the world.

Critical to the expansion of the Registry beyond the confines of the MGH was the arrival of Henrik Malchau, with his vast experience at Goteborg in leading the Swedish National Hip Registry. Dr. Malchau visited the Harris Orthopaedic Lab for a year in 2000 and then returned as a faculty member in 2004, rising to the position of co-director of the Harris Orthopaedic Laboratory and full Professor of Orthopaedic Surgery at Harvard Medical School (HMS). He was subsequently granted the Alan Gerry chair at HMS.

His vision, expertise and experience were critical. As he was formulating the concept of an international organization, the International Society of Arthroplasty Registries, he was living that model by initiating many studies requiring unified data collection from diverse orthopaedic units around the world. This broad, international capacity was critical both in generating large data sets rapidly and encompassing patients as well as surgical techniques from different cultures and societies. Many of these studies also incorporated the valuable technique of radiostereometric analysis (RSA). All the data from these far-flung investigations are managed through the Harris Registry.

One of the most valuable additions to the Registry was the inclusion, automatically, of all germane radiographic images for each patient. Obviously, this occurred after the transition to the use of digital images. The automation grew from the capacity of the Partners HealthCare system to identify specifically each patient in the Registry to enable a daily sweep of all radiographic sites for any new images of each patient for importation into the Registry. Ours was the first registry in the world to do this. Of course, each digital image data provides enhanced value because of the image manipulation that the digital images allow.

Also, in a similar way the clinical records of all of the Partners system are swept daily to acquire any clinical interaction, with each patient at any site in the system.

In addition to participating in the work "Arthroplasty Implant Registries over the Past Five Decades", the Kappa Delta Award (mentioned above), over 150 Harris Orthopaedic Lab papers have had data from the Harris Registry as their backbone. The innovations and advances contained in these papers are detailed in the other chapters of this book.

## SELECTED RELATED REFERENCES

Harris WH. Clinical Results Using the Mueller-Charnley Total Hip Prosthesis. Clin Orthop Rel Res, 86:95-101, 1972.

Harris WH. Preliminary Report of Results of Harris Total Hip Replacement. Clin Orthop Rel Res 95:168-173, 1973.

Stulberg SD, Harris WH. Acetabular Dysplasia and Development of Osteoarthritis of the Hip. In: The Hip, Proceedings of the Second Open Scientific Session of the Hip Society, Ed. WH Harris, St. Louis: CV Mosby Co., pp. 82-93, 1974.

Harris WH. Reconstructive Surgery of the Hip Following Trauma. In: Management of Trauma, Ed. E.F. Cave, J.F. Burke, and R.J. Boyd, Chicago: Year Book Medical Publishers, Inc. pp. 711-717, 1974.

Patel D, Karchmer A, Harris WH. The Role of Preoperative Aspiration of the Hip Prior to Total Hip Replacement. In: The Hip, Proceedings of the Fourth Open Scientific Meeting of the Hip Society. Ed. CM Evarts, St. Louis: CV Mosby, Co., pp. 219-223, 1976.

Harris WH, Crothers OD. Autogenous Bone Grafting Using the Femoral Head to Correct Severe Acetabular Deficiency for Total Hip Replacement. In: The Hip, Proceedings of the Fourth Open Scientific Meeting of the Hip Society. Ed: CM Evarts, St. Louis: CV Mosby Co., pp. 161-185, 1976.

Harris WH, Crothers O, Oh I. Total Hip Replacement and Femoral-Head Bone-Grafting for Severe Acetabular Deficiency in Adults. J. Bone Joint Surg., 59-A:752-759, 1977.

Oh I, Carlson CE, Tomford WW, Harris WH. Improved

Fixation of the Femoral Component after Total Hip Replacement Using a Methacrylate Intermedullary Plug. J. Bone Joint Surg., 60-A: 608-612, 1978.

Harris WH. Total Hip Replacement for Osteoarthritis Secondary to Congenital Dysplasia or Congenital Dislocation of the Hip. International Orthopaedics (SICOT) 2:127-138, 1978.

Harris WH. Loosening. In: The Hip, Proceedings of the Sixth Open Scientific Meeting of the Hip Society. Ed. Carl S Nelson. St. Louis, CV Mosby Co., pp. 162-175, 1978.

Harris WH. Long Stem Femoral Components for Total Hip Replacements. Italian Journal of Orthopedics and Traumatology. Dec. 4:265-277, 1978.

Jupiter JB, Karchmer AW, Lowell JD, Harris WH. Total Hip Arthroplasty in the Treatment of Adult Hips with Current or Quiescent Sepsis. J. Bone and Joint Surg., 63-A: 194-200, 1981.

Harris WH, Allen JR. The Calcar Replacement Femoral Component for Total Hip Arthroplasty: Design, Uses and Surgical Technique. Clin Orthop Rel Res 157:215-224, 1981.

Harris WH. Allografting in Total Hip Arthroplasty: In Adults with Severe Acetabular Deficiency Including a Surgical Technique for Bolting the Graft to the Ilium. Clin Orthop Rel Res 162:150-164, 1982.

Harris WH Jr, White RE. Socket Fixation Using a Metal-Backed Acetabular Component for Total Hip Replacement. A Minimum Five-Year Follow-up. J. Bone Joint Surg., 64-A:745-748, 1982.

Jupiter J, Harris WH. Direct Reimplantation of Total Hip Replacements in Septic Hips in the Adult. Instructional Course Lectures. Vol. 31, CV Mosby, St. Louis: AAOS, pp. 29-38, 1982.

Harris WH, McCarthy JC, O'Neill DA. Femoral Component Loosening Using Contemporary Techniques of Femoral Cement Fixation. J. Bone Joint Surg., 64-A: 1063-1067, 1982.

Harris WH. Revision Surgery for Failed, Nonseptic Arthroplasty. The Femoral Side. Clin Orthop Rel Res 170:8-20, 1982.

Harris WH Jr, McCarthy JC, O'Neill DA. Loosening of the Femoral Component of Total Hip Replacement after Plugging the

Femoral Canal. In: The Hip, Proceedings of the 10th Open Scientific Meeting of the Hip Society, St. Louis: CV Mosby Co., pp.228-238, 1982.

Harris WH Jr., White RE. Resection Arthroplasty for Nonseptic Failure of Total Hip Arthroplasty. Clin Orthop Rel Res, 171:62-67, 1982.

Harris WH Jr, White RE, McCarthy JC, Walker PS, Weinberg EH. Bony Ingrowth Fixation of the Acetabular Component in Canine Hip Joint Arthroplasty. Clin Orthop Rel Res 176:7-11, 1983.

Harris WH, White RE Jr. Advantages of Metal-Backed Acetabular Components for a Total Hip Replacement: A Clinical Assessment with a Minimum 5-Year Follow-up. In: The Hip, Proceedings of the Eleventh Open Scientific Meeting for the Hip Society, CV Mosby Co., St. Louis: Ed. DS Hungerford, Chapter 14, pp.240-246, 1983.

O'Neill DA, Harris WH. Failed Total Hip Replacement Assessment by Plain Radiographs, Arthrograms, and Aspiration of the Hip Joint. J. Bone Joint Surg., 66-A: 540-546, 1984.

Harris WH. Advances in Total Hip Arthroplasty. The Metal-backed Acetabular Component. Clin Orthop Rel Res, 183:4-11, 1984.

Harris WH. Autografting and Allografting in Aseptic Failure of Total Hip Replacement. In: The Hip, Proceedings of the 12th Open Scientific Meeting of the Hip Society, Ed. RB Welch, St. Louis: CV Mosby Co., pp. 286-295, 1984.

McGann W, Mankin HJ, Harris WH. Massive Allografting for Severe Failed Total Hip Replacement. J. Bone Joint Surg., 68-A:4-12, 1986.

Harris WH. Clinical Results of Metal and Plastic Implants in Total Hip Replacement. In: Perspectives on Biomaterials, Ed. OCC Lin and EYS Chao, Elsevier Science Publishers B.V., Amsterdam, pp. 207-221, 1986.

Harris WH, McGann WA. Loosening of the Femoral Component after Use of the Medullary-Plug Cementing Technique.

Follow-up Note with a Minimum Five-year Follow-up. J. Bone Joint Surg., 68-A:1064-1066, 1986.

Gerber SD, Harris WH. Femoral Head Autografting to Augment Acetabular Deficiency in Patients Requiring Total Hip Replacement. J. Bone Joint Surg., 68-A: 1241-1248, 1986.

Harris WH. Bone grafting for acetabular deficiency in association with total replacement. In: The Hip, Proceedings of the Fourteenth Open Scientific Meeting of the Hip Society, CV Mosby Co., Ed. RA Brand, St. Louis: pp. 94-119, 1986.

Jasty M., Harris WH. Total Hip Reconstruction Using Frozen Femoral Head Allografts in Patients with Acetabular Bone Loss. Orthop. Clin. of N. Am. 18:291-299, 1987.

Harris WH, Penenberg BL. Further Follow-up on Socket Fixation using a Metal-Backed Acetabular Component for Total Hip Replacement. A Minimum Ten-year Follow-up Study. J. Bone and Joint Surg. 69-A:1140-1143, 1987.

Schutzer S, Harris WH. Trochanteric Osteotomy for Revision Total Hip Arthroplasty. 97% Union Rate Using a Comprehensive Approach. Clin Orthop Rel Res 227:172-183, 1988.

Schutzer SF, Harris WH. Deep-wound infection after total hip replacement under contemporary aseptic conditions. J. Bone Joint Surg. 70-A:724-727, 1988.

Harris WH, Krushell RJ, Galante JO. Results of Cementless Revisions of Total Hip Arthroplasties Using the Harris-Galante Prosthesis. Clin Orthop Rel Res 235:120-126, 1988.

Rubash HE, Harris WH. Revision of Nonseptic, Loose, Cemented Femoral Components Using Modern Cementing Techniques. J. Arthroplasty 3:241-248, 1988.

Jasty M, Harris WH. Results of Total Hip Reconstruction Using Acetabular Mesh in Patients with Central Acetabular Deficiency. Clin Orthop Rel Res 237:142-149, 1988.

Davey JR, Harris WH. A Preliminary Report of the Use of a Cementless Acetabular Component with a Cemented Femoral Component. Clin Orthop Rel Res 245:150-155, 1989.

Harris WH, Maloney WH. Hybrid total hip arthroplasty. Clin Orthop Rel Res 249:21-29, 1989.

Jasty M, Harris WH. Experience with Cementless Porous-Surfaced Acetabular Components. Orthop. Rel. Sci. 1:52-61, 1990.

Jasty M, Harris WH. Salvage total hip reconstruction in total hip reconstruction in patients with major acetabular bone deficiency using structural femoral head allografts. J. Bone Joint Surg. 72-B: 63-67, 1990.

Johnston RC, Fitzgerald RH, Harris WH, Poss R, Muller ME. Clinical and Radiographic Evaluation of Total Hip Replacement. A Standard System of Terminology for Reporting Results. J. Bone Joint Surg. (Am) 72-A: 161-168, 1990.

Mulroy RD Jr., Harris WH. Failure of acetabular autogenous grafts in total hip arthroplasty. Increasing incidence: a follow-up note. J. Bone Joint Surg., 72-A: 1990, 1536-1540.

Mulroy RD Jr, Mankin HJ, Harris WH. Insertion of a Prosthetic Hip into a Total Hip Allograft. J. Bone Joint Surg. 72-B: 4; 643-646;1990.

Maloney WJ, Jasty M, Harris WH, Galante JO, Callaghan JJ. Endosteal Erosion in Association with Stable Uncemented Femoral Components. J Bone Joint Surg. 72-A: No. 7, 1025-1034, 1990.

Maloney WJ, Harris WH. Comparison of a Hybrid with an Uncemented Total Hip Replacement. J. Bone Joint Surg. (Am) 72-A: 1349-1352, 1990.

Maloney WJ, Jasty M, Rosenberg A, Harris WH. Bone Lysis in Well-Fixed Cemented Femoral Components. J. Bone Joint Surg. Br. 72-B: 970, 1990.

Mulroy RD Jr, Harris WH. Failure of Acetabular Autogenous Grafts in Total Hip Arthroplasty. Increasing Incidence: A Follow-Up Note. J Bone Joint Surg. (Am) 72-A: 1536-1540, 1990.

Maloney WJ, Krushell RD, Jasty M, Harris WH. Incidence of Heterotopic Ossification after Total Hip Replacement: Effect of the Type of Fixation of the Femoral Component. J. Bone Joint Surg. (Am) 73-A: 191-193, 1991.

Russotti GM, Harris WH. Proximal Placement of the Acetabular

Component in Total Hip Arthroplasty. A Long-Term Study. J Bone Joint Surg., 73-A: 587-592, 1991.

Jasty M, Maloney WJ, Harris WH, Galante JO, Callaghan JJ. Endosteal osteolysis around well-fixed porous coated cementless femoral components. In: ASTP STP 114, K.R. St. John, Ed., Am. Soc. for Testing and Materials, Philadelphia, PA, pp. 61-67, 1992.

Tanzer M, Maloney WJ, Jasty M, Harris WH. The progression of femoral cortical osteolysis in association with total hip arthroplasty without cement. J.Bone Joint Surg. 74-A: (3):404-410, 1992.

Barrack RL Jr, Mulroy RD Jr, Harris WH. Improved Cementing Techniques and Femoral Component Loosening in Young Patients with Hip Arthroplasty: a 12 year radiographic review. J Bone Joint Surg. 74-B: 385-389, 1992.

Barrack RL, Jasty M, Bragdon CR, Haire T, Harris WH. Thigh Pain Despite Bone Ingrowth into Uncemented Femoral Stems. J Bone Joint Surg. 74-B: 507-510, 1992.

Maloney WJ, Jasty M, Willett C Jr, Mulroy RD Jr, Harris WH. Prophylaxis for Heterotopic Bone Formation after Total Hip Arthroplasty Using Low Dose Radiation in High Risk Patients. Clin Orthop Rel Res 280:230-234, 1992.

Jasty M, Mulroy RD Jr, Harris WH. Total Hip Replacement and Implant Interfaces. Ch. 26. p. 276-288. In: Habala MB, Reddi AH. Eds.: Bone Grafts and Bone Substitutes. WB Saunders Co., Philadelphia, 1992.

Tanzer M, Drucker D, Jasty M, McDonald M, Harris WH. Revision of the Acetabular Components with an Uncemented Harris-Galante Porous-Coated Prosthesis. J. Bone Joint Surg. 74-A: 987-994, 1992.

Schmalzried TP, Harris WH. The Harris-Galante Porous-Coated Acetabular Component with Screw Fixation. Radiographic Analysis of Eighty-Three Primary Hip Replacements. J Bone Joint Surg. 74-A:1130-1139, 1992.

Harris WH. Is it Advantageous to Strengthen the Cement-Metal Interface and Use a Collar for Cemented Femoral Components of Total Hip Replacements? Clin Orthop Rel Res 285:67-72, 1992.

Jasty M, Jiranek W, Harris WH. Acrylic Fragmentation in Total Hip Replacements and its Biological Consequences. Clin Orthop Rel Res 285:116-128, 1992.

Barrack RL, Harris WH. The Value of Aspiration of the Hip Joint Before Revision Total Hip Arthroplasty. J. Bone Joint Surg. 75-A: 66-76, 1993.

Maloney WJ, Davey JR, Harris WH. Bead Loosening from a Porous-Coated Acetabular Component: a follow-up note. Clin Orthop Rel Res 281:112-114, 1992.

Harris WH. Bulk versus Morselized Bone Graft in Acetabular Revision Total Hip Replacement. Seminars in Arthroplasty 4:68-71, 1993.

Goetz DD, Harris WH. Why Have We Left Charnley Low Friction Arthroplasty? Iowa Orthop. J. 13:29-39, 1993.

Schmalzried TP, Harris WH. Hybrid total hip replacement. A 6.5 year follow-up study. J. Bone Joint Surg. Br, 75-B: 608-615, 1993.

Kwong LM, Jasty M, Harris WH. High failure rate of bulk femoral head allografts in total hip acetabular reconstructions at 10 years. J. Arthroplasty 8:341-346, 1993.

Barrack RL, Burke DW, Cook SD, Skinner HB, Harris WH. Complications Related to Modularity of Total Hip Components. J. Bone Joint Surg. 75-B: 688-692, 1993.

Harris WH. "Hybrid Total Hip Replacement." Ch. 33 in Joint Replacement. State of the Art, Coombs R, Gristina A, Hungerford D, Eds. Mosby Year Book, St. Louis, MO, 1990.

Harris WH. Management of the Deficient Acetabulum Using Cementless Fixation without Bone Grafting. Orthop. Clin. N. Amer. 24:663-665, 1993.

Harris WH. Total Hip Replacement in the Middle-Aged Patient. Contemporary Cementing for Fixation of the Femoral Component. Orthop. Clin N Amer 24:611-616, 1993.

Estok DM II, Harris WH. Long-term Results of Cemented Femoral Revision Surgery Using Second-Generation Techniques. An Average 11.7 Year Follow-Up Evaluation. Clin Orthop Rel Res 299:190-202, 1994.

Pierson JL, Harris WH. Cemented Revision for Femoral Osteolysis in Cemented Arthroplasties. J. Bone Joint Surg. Br. 76-B: 40-44, 1994.

Harris WH. The Case for Cemented Fixation of the Femur in Every Patient. Ch. 36. In: Instructional Course Lectures, Vol. 43, M Schafer, Ed. p. 367-371. AAOS: Rosemont IL, 1994.

Harris WH. Osteolysis and Particle Disease in Hip Replacement. A Review. Acta Orthop. Scand. 65: 113-124, 1994.

Schmalzried TP, Wessinger SJ, Hill GE, Harris WH. The Harris-Galante Porous Acetabular Component Press-fit without Screw Fixation. J. Arthroplasty 9:235-242, 1994.

Schutzer SF, Harris WH. High placement of porous-coated acetabular components in complex total hip arthroplasty. J. Arthroplasty. 9(4):359-67, 1994 Aug.

Goetz DD, Smith EJ, Harris WH. The prevalence of femoral osteolysis associated with components inserted with and without cement in total hip replacements. J Bone Joint Surg. 76-A :(8)1121-1129, 1994.

Harris WH. Total Hip Replacement: "Cement versus Cementless" Resolution. Bulletin on the Rheumatic Diseases 43:1-4, 1994.

Jasty M, Bragdon CR, Jiranek W, Chandler H, Maloney W, Harris WH. Etiology of osteolysis around porous-coated cementless total hip arthroplasties. Clin Orthop Rel Res 308:111-126, 1994.

Harris WH. Improved Long-Term Results of Femoral Fixation in Revision Surgery Using Modern Cementing. Ch. 31, pg. 289-294. In: Total Hip Revision Surgery. Galante JO, Rosenberg AG, Callaghan JJ, Eds. Bristol-Myers-Squibb/Zimmer Orthopaedic Symposium Series. New York, Raven Press: 1995.

Jasty M, Harris WH. Cementless Acetabular Revisions. Ch. 35. pg. 317-323. In: Total Hip Revision Surgery. Galante JO, Rosenberg AG, Callaghan JJ Eds. Bristol-Myers-Squibb/Zimmer Orthopaedic Symposium Series. New York, Raven Press: 1995.

Harris WH. The Problem is Osteolysis. Clin Orthop Rel Res 311:46-53, 1995.

Jasty M, Anderson MJ, Harris WH. Total Hip Replacement for

Developmental Dysplasia of the Hip. Clin Orthop Rel Res 311:40-45, 1995.

Smith E, Harris WH. Increasing prevalence of femoral lysis in cementless total hip arthroplasty. J. Arthroplasty 10:407-412, 1995.

Pierson JL, Harris WH. Effect of improved cementing techniques on the longevity of fixation in revision cemented femoral arthroplasties: Average 8.8 year follow-up period. J. Arthroplasty 10:581-591, 1995.

Harris WH. The Lessons of Acetabular Component Fixation. Acetabular Component Fixation in the 1990's. Orthopedics, 18 (9):811-812, 1995.

Harris WH. Cemented Revision for Femoral Failure. Revision Total Hip Arthroplasty. Orthopedics, 18(9):854-855, 1995.

Mulroy WF II, Estok DM, Harris WH: Total hip arthroplasty with use of so-called second-generation cementing techniques. A fifteen-year-average follow-up study. J. Bone Joint Surg. 77-A (12) 1845-1852, 1995.

Schutzer SF, Grady-Benson J, Jasty M, O'Connor DO, Bragdon CR, Harris WH. Influence of Intraoperative Femoral Fractures and Cerclage Wiring on Bone Ingrowth into Canine Porous-Coated Femoral Components. J. Arthroplasty 10 (6): 823-829, 1995.

McLaughlin JR, Harris WH. Revision of the femoral component of a total hip arthroplasty with the calcar-replacement femoral component. Results after a mean of 10.8 years postoperatively. J Bone Joint Surg. 78-A:331-339, 1996.

Jasty M, Estok DM, Harris WH. The mechanisms involved in the failure of fixation of components in total hip arthroplasty. Seminars in Arthroplasty 7:76-85, 1996.

Mulroy WF, Harris WH. Revision total hip arthroplasty with use of so-called second-generation cementing techniques for aseptic loosening of the femoral component. A fifteen year follow-up study. J Bone Joint Surg. 78-A:325-330, 1996.

Harris WH, Smith EJ, Goetz DD. Bone Cement as a Seal Protecting the Femur from the Ingress of Particulate Wear Debris.

Instructional Course Lectures, Douglas J Pritchard, MD, Editor, Vol. 45. AAOS pp. 183-185, Rosemont, IL, 1996.

Schmalzried TP, Peters PC, Maurer BT, Bragdon CR, Harris WH. Long-duration Metal on Metal Total Hip Arthroplasties with Low Wear of the Articulating Surfaces. J. Arthroplasty 11:322-331, 1996.

Harris WH. Hybrid total hip replacement: Rationale and intermediate clinical results. Clin Orthop Rel Res 333:115-164, 1996.

Maloney WJ, Sychterz C, Bragdon CR, McGovern T, Jasty M, Engh CA, Harris WH. The Otto Aufranc Award. Skeletal response to well fixed femoral components inserted with and without cement. Clin Orthop Rel Res 333:15-26, 1996.

Shinar AA, Harris WH. Bulk structural grafts and allografts for reconstruction of the acetabular in total hip arthroplasty. Sixteen-year average follow-up. J Bone Joint Surg. Am. 79A:159-168, 1997.

Jasty M, Goetz DD, Lee KR, Hanson AE, Elder JR, Harris WH. Wear of polyethylene acetabular components in total hip arthroplasty. An analysis of 128 components retrieved at autopsy or revision operation. J Bone Joint Surg. 79-A: 349-358, 1997.

Mulroy W, Harris WH. Acetabular and femoral fixation 15 years after cemented total hip surgery. Clin Orthop Rel Res 337:118-128, 1997.

Harris WH. Options for primary femoral fixation in total hip arthroplasty. Cemented stems for all. Clin Orthop Rel Res 344:118-123, 1997.

Smith SE, Harris WH. Total hip arthroplasty performed with the insertion of a femoral component with cement and the acetabular component without cement. Ten to thirteen year results. J Bone Joint Surg. 79-A:1827-1833, 1997.

Bal BS, Maurer BT, Harris WH. Trochanteric union following revision total hip arthroplasty. J. Arthroplasty 13:29-33, 1998.

Smith SE II, Estok DM, Harris WH. Average 12-year outcome of a chrome-cobalt, beaded, bony ingrowth acetabular component. J. Arthroplasty 13:50-60, 1998.

Smith SE II, Harris WH. The hybrid total hip replacement.

Chapter 13. In: Total Hip Arthroplasty Outcomes. Ed. by Finerman, GAM, Dorey FJ, Grigoris P, McKellop HA, p. 215-225. Churchill Livingstone, NY: 1997.

Harris WH. Total hip arthroplasty in the management of congenital hip dislocation. Ch. 73. In: The Adult Hip. Ed. Callaghan JJ, Rosenberg AG, Rubash HE. Lippincott-Raven Publishers, Philadelphia, 1998.

Shinar AA, Harris WH. Cemented total hip arthroplasty following previous femoral osteotomy: An average 16-year follow up study. J. Arthroplasty 13: 243-253, 1998.

Harris WH. "The Hybrid Total Hip Replacement." Ch. 13. In: The Hip. Editor: Sledge C. pp.239-258, Lippincott Raven, NY, 1998.

Bal B, Vandelone D, Gurba D, Jasty M, Harris WH. Polyethylene wear in cases using femoral stems of similar geometry but different metals, porous layer and modularity. J. Arthroplasty pp. 492-499, 1998.

Harris WH. Revision Total Hip Arthroplasty: Reconstruction at a High Hip Center in Acetabular Revision Surgery Using a Cementless Acetabular Component. Orthopedics. 12:9: 991-992, 1998.

Dearborn JT, Harris WH. Postoperative Mortality after Total Hip Arthroplasty: An Analysis of Deaths after 2,736 Procedures. J Bone Joint Surg. 80-A(9): 1291-1294, 1998.

Harris WH. Long Term Results of Cemented Femoral Stems with Roughened Precoated Surfaces. Clin Orthop Rel Res 355:137-143, 1998.

Smith S II, Estok DM, Harris WH. Total Hip Arthroplasty with Use of Second- Generation Cementing Techniques: An Eighteen-Year-Average Follow-up Study. J Bone Joint Surg. 80-A (11):1632-1640, 1998.

Clohisy JC, Harris WH. The Harris-Galante Porous-Coated Acetabular Component with Screw Fixation: An Average 10-Year Follow-Up Study. J Bone Joint Surg. 81-A (1): 66-73, 1999.

Clohisy JC, Harris WH. Primary Hybrid Total Hip Replacement,

Performed with Insertion of the Acetabular Component without Cement and a Precoat Femoral Component with Cement: An Average ten-year follow-up study. J Bone Joint Surg. 81-A (2):247-255, 1999.

Anderson MJ, Harris WH. Total Hip Arthroplasty with Insertion of the Acetabular Component without Cement in Hips with Total Congenital Dislocation or Marked Congenital Dysplasia. J Bone Joint Surg. 81-A (3):347-54, 1999.

Dearborn JT, Harris WH. High Placement of an Acetabular Component Inserted without Cement in a Revision Total Hip Arthroplasty: Results after a Mean of Ten-Years. J Bone Joint Surg. 81-A (4):469-480, 1999.

Bal BS, Maurer T, Harris WH. Revision of the Acetabular Component without Cement after a Previous Acetabular Reconstruction with Use of a Bulk Femoral Head Graft in Patients who had Congenital Dislocation or Dysplasia: A Follow-up Note. J Bone Joint Surg. 81-A:1703-1706, 1999.

Maloney WJ, Anderson M, Harris WH, Kraay M, Rubash HE, Woolson ST. Acetabular Fixation in Primary Total Hip Arthroplasty: Fixation, Polyethylene Wear, and Pelvic Osteolysis in Primary Total Hip Replacement. Clin Orthop Rel Res 369:157-164, 1999.

Schmalzried TP, Brown IC, Amstutz HC, Engh CA, Harris WH. The Role of Acetabular Component Screw Holes and/or Screws in the Development of Osteolysis. Proc. Inst. Mech. Eng. [H], 213 (2):147-53, 1999.

Clohisy JC, Harris WH. The Harris-Galante Uncemented Femoral Component in Primary Total Hip Replacement at 10-years. J Arthroplasty 14:8, 915-917, 1999.

Dearborn JT, Harris WH. Acetabular Revision Arthroplasty using So-Called Cementless Components: An Average 7-Year Follow-up. J Arthroplasty (15) 8-15, 2000.

Smith SE II, Estok DM, Harris WH. 20-Year Experience with Cemented Primary and Conversion Total Hip Arthroplasty using So-Called Second-Generation Cementing Techniques in Patients Aged 50 Years of Younger. J Arthroplasty 15:3, 263-273, 2000.

Dearborn JT, Harris WH. Acetabular Revision after Failed Total Hip Arthroplasty in Patients with Congenital Hip Dislocation and Dysplasia: Results After a Mean of 8.6 Years. J Bone Joint Surg. 82-A: (8), 1146-1153, 2000.

Clohisy JC, Harris WH. Matched-Pair Analysis of Cemented and Cementless Acetabular Reconstruction in Primary Total Hip Arthroplasty. J Arthroplasty 16(6):697-705, 2001.

Behairy Y, Meldrum RD, Harris WH. Hybrid Revision Total Hip Arthroplasty: A 7-Year Follow-Up Study. J. Arthroplasty, 16(7), 829-837, 2001.

Hauser DL, Wessinger SJ, Condon RT, Golladay GJ, Hoeffel DP, Gillis DJ, Merrill DR, Chaisson D, Freiberg AA II, Estok DM, Rubash HE, Malchau H, Harris WH. An Electronic Database for Outcome Studies that Includes Digital Radiographs. J. Arthroplasty 16(8) Suppl. 1, 71-75, 2001.

Behairy Y, Harris WH. Mode of Loosening of Matte-Finished Femoral Stems in Primary Total Hip Replacement. Saudi Med. J. 23:10, 1187-94, 2002.

Maloney WJ, Schmalzried T, Harris WH. Analysis of long-term cemented total hip arthroplasty retrievals. Clin Orthop Rel Res 405:70-78, 2002.

Harris WH. Results of Uncemented Cups: A Critical Appraisal at 15 Years. Clin Orthop Rel Res 417:121-125, 2003.

Harris WH. The three revolutions in acetabular revision surgery for total hip replacement: 1. Definite and 2. Probable. Chirurgia Degli di Movimento. 88(1): 1-13, 2003.

Jamali A, Dungy D, Mark A, Schule S, Harris WH. Isolated Acetabular Revision with Use of the Harris-Galante Cementless Component: Study with Intermediate Term Follow Up. J. Bone Joint Surg.86-A (8):1690-1697, August 2004.

Hallstrom R. Golladay GJ, Vittetoe DA, Harris WH. Cementless acetabular revision with the Harris-Galante porous prosthesis. Results after a minimum of ten years of follow-up. J. Bone Joint Surg. 86-A (5): 1007-1011, 2004.

Manning D, Chiang PP, Martell JM, Galante JO, Harris WH.

In Vivo Comparative Wear Study of Traditional and Highly Cross-linked Polyethylene in Total Hip Arthroplasty. J. Arthroplasty Oct. 20(7): 880-886, 2005.

Hendricks KJ, Harris WH. Revision of failed acetabular components with use of so-called jumbo non-cemented components. A concise follow-up of a previous report. J. Bone Joint Surg. Mar 88-A(3):559-63, 2006.

Geller JA, Malchau H, Bragdon C, Greene M, Harris WH, Freiberg AA. Large diameter femoral heads on highly cross-linked polyethylene: minimum 3-year results. Clin Orthop Rel Res Jun 447:53-59, 2006.

Hampton BJ, Harris WH. Primary Cementless Acetabular Components in Hips with Severe Dysplasia or Total Dislocation. A Concise Follow-Up, At an Average of Sixteen Years, of a Previous Report. J Bone Joint Surg.; 88-A (7); p. 1549-1552: 2006 July.

Hendricks KJ, Harris WH. High placement of noncemented acetabular components in revision total hip arthroplasty. A concise follow-up, at a minimum of fifteen years, of a previous report. J.Bone Joint Surg. Oct. 88-A(10):2231-2236, 2006.

Bragdon CR, Barrett S, Martell JM, Greene ME, Malchau H, Harris WH. Steady-State Penetration Rates of Electron Beam-Irradiated, Highly Crosslinked Polyethylene at an Average 45-Month Follow-Up. J. Arthroplasty; Oct. 21(7): 935-943, 2006.

Harris WH. An Integrated Solution of Acetabular Revision Surgery. Clin Orthop Rel Res Dec. 453: 178-182, 2006.

Bragdon CR, Kwon YM, Geller JA, Greene ME, Freiberg AA, Harris WH, Malchau H. Minimum 6 Year Followup of Highly Crosslinked Polyethylene THA. Clin Orthop Rel Res; Dec. 465, 122-127, 2007.

Bragdon C, Doerner M, Rubash H, Kwon YM, Martell J, Clohisy J, White R, Della Valle C, Berry D, Jarrett B, Lachiewicz P, Bertin K, Johanson P, Palm H., Harris W, Malchau H. Clinical Multi-centric Studies of the Wear Performance of Highly Crosslinked Remelted Polyethylene in THR. Clin Orthop Rel Res 471(2): 393-402, 2013.

Hauser DL, Wessinger SJ, Condon RT, Golladay GJ, Hoeffel DP, Gillis DJ, Merrill DR, Chaisson D, Freiberg II AA, Estok DM, Rubash HE, Malchau H, Harris WH. An Electronic Database for Outcome Studies that Includes Digital Radiographs. J. Arthroplasty 16(8) Suppl. 1, 71-75, 2001.

Malchau H, Garellick G. Arthroplasty Implant Registries Over the Past Five Decades, Development, Current and Future Impact. Kappa Delta and Orthopedic Research and Education Foundation (OREF) Clinical Research Awards, San Diego CA, March 2017.

# ACKNOWLEDGEMENTS

CLEARLY A FULL REPETITION OF the Dedication of this book constitutes the dominant portion of the acknowledgements for its creation.

Special in addition are the many members of this medical and medical research community who have inspired and aided our adventures in creating the Harris Orthopaedic Laboratory and its productivity before and after its founding in 1969. Key among the many are two, Edith Weinberg, a remarkable compatriot in the trenches over all these years without whom the Harris Orthopaedic Laboratory would not exist, and Melvin J. Glimcher, the "one of a kind" creator of serious contemporary research at the MGH.

Unique in its role as the originator of both momentum and flexibility for our work is the paramount role of the William H. Harris MD Foundation, founded in 1968, which was and is crucial to sustaining the funding that was essential for these many worldwide medical advances.

Specifically for the creation of this book two people warrant recognition, my third son David Daniel Harris, who did the essential exploration of the mysteries of self publication that underlie this book and Anne Goodrich, who endlessly and patiently converted my illegible script into what you see today. To both, very special thanks.

# INDEX

www.ingramcontent.com/pod-product-compliance
Lightning Source LLC
Chambersburg PA
CBHW030003190526
45157CB00014B/409